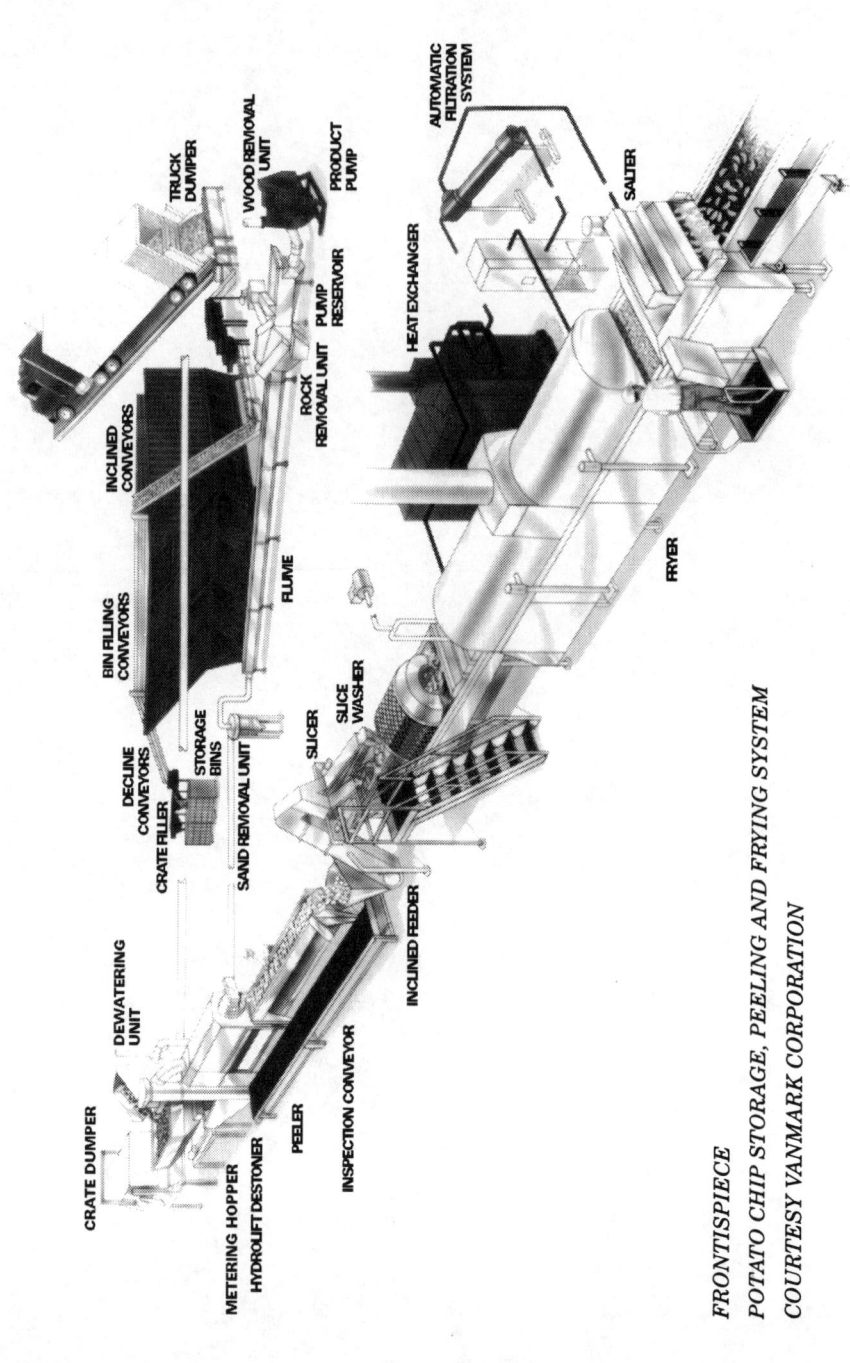

FRONTISPIECE
POTATO CHIP STORAGE, PEELING AND FRYING SYSTEM
COURTESY VANMARK CORPORATION

UNIT OPERATIONS FOR THE FOOD INDUSTRIES

By

Wilbur A. Gould, Ph.D.

*Emeritus Professor,
Food Processing and Technology,
The Ohio State University
and
Executive Director,
Mid-America Food Processors Association
and
Consultant to the Food Industries*

UNIT OPERATIONS FOR THE FOOD INDUSTRIES

A technical reference book and textbook for students of food technology, food plant managers, product research and development specialists, food brokers, technical salesmen, food equipment manufacturers, and food industry suppliers.

COPYRIGHT ©1996

CTI PUBLICATIONS, INC.
all rights reserved

ISBN Numbers are as follows: 0-930027-29-9

Library of Congress Catalog-in-Publication Data

Gould, Wilbur A., 1920-
 Unit Operations For The Food Industries, by Wilbur A. Gould.
 p.   cm.
 Includes Index
 ISBN 0-930027-29-9 (hardbound)
  1. Food Industry and trade — Automation.   I. Title.
TP372.8.G686   1996
664'.02 — dc20   95-48455
                                                    CIP

No part of this book may be reproduced in any form or by any means—graphic, electronic, or mechanical, including photocopying, recording, taping, or information storage and retrieval system, without written permission from the publishers.

While the recommendations in this publication are based on scientific studies and industry experience, references to basic principles, operating procedures and methods, types of instruments and equipment, and food formulas, are not to be construed as a guarantee that they are sufficient to prevent damage, spoilage, loss, accidents or injuries, resulting from use of this information. Furthermore, the study and use of this publication by any person or company is not to be considered as assurance that that person or company is proficient in the operations and procedures discussed in this publication. The use of the statements, recommendations, or suggestions contained, herein, is not to be considered as creating any responsibility for damage, spoilage, loss, accident or injury, resulting from such use.

COVER PHOTO
CORN AND TORTILLA CHIP PROCESSING AND PACKAGING SYSTEM
COURTESY HEAT AND CONTROL, INC.

A PUBLICATION OF

CTI Publications, Inc.
2 Oakway Road, Timonium, Maryland 21093-4247 USA

PREFACE

Unit Operations has been a subject of great interest to me over the years. It is the subject of research, on my part, to better understand the parameters involved to control product quality and to process food efficiently and safely. Unit Operations has been a large part of my life in terms of teaching, both at the academic level and within the industry, in my workshops and consulting efforts. I do not pretend to have all the answers, as I continue to learn from my colleagues and peers within this great industry. Truthfully, I am greatly amazed by the knowledge and know-how which already exists.

Unit operations have changed drastically over the years, with the newer technologies really making a great impact on the industry and the resulting finished products. The changes in unit operations have produced a change from hand labor to automation. The adoption of modern technology has increased efficiency and productivity within the factory. Most importantly, utilizing the newer unit operations has greatly improved product quality.

The increased development of newer unit operations will continue to improve operating conditions within the factory and improve finished product qualities. The only limitations will be the operator, in learning to utilize these new and improved technologies and to put them into practice.

Plant and unit operations operators need to take advantage of the updated and improved changes in equipment. They need to observe these changes and know what is available to help them help themselves. Management must provide the facts and use these facts and other information to justify these changes as they become available.

Most importantly, students need to understand unit operations and learn the parameters and controls of their operations. Ideally, they need to get hands-on experience with modern unit operations to better control and assure product quality.

This field is more fascinating than any of the other many aspects of the food business, and by the continued dedication of the manufacturers, distributors, and suppliers, we all can benefit as new technologies continue to come forth. My only hope is that all those who are exposed to this aspect of the food business will become enthused and eager to put into practice the various unit operation know-how's which are available to them.

— **Wilbur A. Gould**

ACKNOWLEDGEMENT

I am deeply indebted to my many friends and associates in the food industry for their excellent help, their full dedication, and their complete cooperation in letting me put this material together in a book. This book could not and should not cover every aspect of all the numerous unit operations in the food industries. However, it is a start and does cover a number of unit operations that should be helpful to all that work in this industry. One hesitates to thank individuals for fear of leaving out some who are most important; however, the following persons and their firms have overwhelmed me with their interest in this subject and their sharing of some most important information. With that in mind, my thanks to these persons and their firms:

- Marvin Gerdes, FMC Corporation
- Bob Green, Buckeye Pumps, Inc.
- Don Giles, Heat and Control, Inc.
- Jeff Lamb, Vanmark Corporation
- Jim Townsend, APV Crepaco, Inc.
- Dave Olney, G. J. Olney, Inc.
- Charles Leader, Leader Engineering
- and AK Robins, Inc.

I also wish to sincerely thank the late H. D. Brown, who taught me the principles of observation and knowledge of testing; my former graduate students who helped me, through their studies, better understand specific unit operations; and to my friends in the food processing industry for letting me observe and study their factories. Lastly, my thanks to Randy Gerstmyer and Art Judge, II, CTI Publications, for encouraging me in this endeavor.

This industry has been great to me, and this book was put together to say "thank you" for helping me have a better understanding of the parts you all have within this industry. I share my information in the hopes of making others aware of aspects of this great industry to help them seek more information. Knowledge is a real blessing and a reward to those who continually seek it.

— **Wilbur A. Gould**

THIS BOOK BELONGS TO:

_____

CONTENTS

UNIT OPERATIONS FOR THE FOOD INDUSTRIES

	Preface	v
	Acknowledgement	vi
Chapter 1.	Introduction	1
Chapter 2.	Materials Handling Including Receiving	11
Chapter 3.	Cleaning	27
Chapter 4.	Quality Separation	33
Chapter 5.	Peeling	41
Chapter 6.	Disintegration—Little Change in Form	49
Chapter 7.	Disintegration—Considerable Change in Form	57
Chapter 8.	Separating	65
Chapter 9.	In-Line Protective Equipment	69
Chapter 10.	Blanching, Scalding and Precooking	75
Chapter 11.	Pumps and Pumping	79
Chapter 12.	Mixing and Blending	89
Chapter 13.	Salting and Brining, Sugars and Syrups, Seasoning, Enrobing, Batter and Breading	97
Chapter 14.	Exhausting and Mechanical Vacuumizing	107
Chapter 15.	Filling	111
Chapter 16.	Packaging, Sealing and/or Closure and Coding	119
Chapter 17.	Canning and Thermal Sterilization	125
Chapter 18.	Freezing	137
Chapter 19.	Drying and Dehydration	143
Chapter 20.	Frying	151
Chapter 21.	Extrusion Cooking	159
Chapter 22.	Assuring The Safety Of Our Food	163
Chapter 23.	Maintenance, Repairs and People	165
	Appendix	166
	References and Suggested Additional Readings	168
	Index	171

Chapter 1

INTRODUCTION

AN INDUSTRY OVERVIEW

Today, some 90% of all the food we eat is processed in one form or another. The food processing industries employ one out of every seven working persons. The food industry is a multi-billion dollar industry and growing in most categories.

The canning of food started in the mid-1800's and has progressed to the point today where each consumer uses some 500 cans or packages of food per year. It is a process whereby food is either heated aseptically to destroy organisms of public health significance prior to aseptically filling and container closure, or the food is filled in the container, sealed and cooked to destroy organisms of public health significance. The processed canned products have a shelf life of approximately 30 months for most items.

Frozen foods are relatively new, being first commercially processed around 1930. The food is prepared as for canning, but frozen in the package or individually quick freezing (IQF) of each piece of food and then packaged and held at 0°F until ready for consumption. Much of the frozen food today is accomplished as a secondary procedure, that is, the processor manufactures and packages complete dinner entrees using the individual frozen items as portions of the entree. In either case, the product has a shelf life of approximately 1 year.

Dried foods are considered one of the oldest forms of food preservation. In Biblical times, foods were sun dried and stored for later uses. Today, the industry still dries foods using solar energy; however, most of the commercially dried foods are dried under conditions of controlled temperature and relative humidity (commonly called dehydrated). Dried or dehydrated foods today constitute a major portion of man's food, including the vast amounts of flour, sugars, salt, spices, cereals, fruits and many prepared products, like potatoes and other root crops, and other vegetables. Most dried foods have a shelf life in excess of 2 years.

Unit Operations are the many steps in any food processing operation. This book is all about some of these steps, that is, the unit operations involved in the preparation and preservation of many of the major food commodities. As Burton, Cruess, Brown, Parker and others have earlier described, "A unit operation is usually considered as a physical step in processing or manufacture that is incapable of division into smaller units."

In this book, our focus is primarily on those physical unit operations of particular importance in the processing of the major fruits, vegetables and related products preserved by canning, freezing, drying, frying, and other forms of manufacture.

These unit operations include the following: Receiving, Materials Handling; Cleaning–Wet and Dry; Quality Separation; Peeling; Disintegration with little change in form, i.e., Husking, Pitting, Coring, Shelling, Snipping, etc.; Disintegration with considerable change in form, i.e., Cutting, Shredding, Crushing, Comminuting, Homogenizing, Sheeting, and Extracting/Juicing, etc.; Separating by Screening, Floating, Settling, Filtering, and Deaeration; Concentration; Blanching; Pumping; Mixing and Blending; Exhausting and Vacuuming; Salting, Seasoning, Enrobing, Batter and Breading; Filling, Sealing and Closure, and Packaging, including Coding; Extrusion; Preservation, i.e., Canning, Freezing, Drying, and Frying.

SOME INDUSTRY CHARACTERISTICS

The food industries are going through major changes, with the larger firms becoming larger and the smaller firms becoming fewer in numbers. Mergers are still most active, with many of the smaller inefficient facilities being closed and their technology utilized by the larger firms.

The industry is still producing more processed food today than ever before and with greater productivity and the food is generally of much higher qualities. Efficiency is being improved and many new products enter the market every day.

The food industry is usually organized into the following basic areas:

(1) Procurement and Receiving of Raw Materials,
(2) Preparation and Packaging,
(3) Processing and Warehousing,
(4) Quality Assurance,

INTRODUCTION

(5) Marketing and Sales, and, of course,
(6) Management (see Table 1.1).

The money for operating the business is primarily provided by and/or through the Board of Directors. There are still many U.S. family-owned businesses today that are in the 4th and 5th generation and, of course, there are many, many large stock corporations.

The following generic Flow Chart (Figure 1.1, page 4) sets forth, as an example, the large number of unit operations. One must understand that there are many significant differences in this flow chart, depending on the commodity being processed. Also, one must understand that there are significant differences in the method of food preservation, depending on whether it is dried, fried, frozen or canned.

COMPARISON OF INDUSTRIES

Freezing and drying of food are a much simpler process when compared to the technology utilized in canning. However, freezing, canning and drying of foods all require the same type of preparation with the change starting at the preservation and packaging area. Unless asepticly processed, canning requires filling the containers and covering with brine or sugar solutions, hermetically sealing the containers, and cooking and cooling the product in the can.

Freezing is usually accomplished by freezing each unit of the product (IQF) on conveyor belts and then the product is packaged and stored at 0 degree F. until used by the customer.

Drying and frying are similar to freezing in that the units are dried or fryed before packaging. The advantage of dried foods is that they are shelf stable and do not need to be refrigerated or frozen. Their shelf life may be longer than their canned or frozen counterparts. Fried foods may have a relatively short shelf life, due in great part to the oils used in preserving the product, as they may go rancid rather quickly. Shelf life can be prolonged by reducing the oil content and by packaging the product in an inert gas in a moisture vapor proof container.

All processes produce essentially the same finished product as far as technology is concerned. Some customers prefer snacks or frozen products, while others prefer the dried or canned product. It should be pointed out that tomatoes and some other fruits are hardly ever frozen, seldom dried, and hardly ever fried.

The canning industry has developed many multiple uses for some products like tomatoes, that is, various styles of peeled tomatoes,

4 UNIT OPERATIONS FOR THE FOOD INDUSTRIES

FIGURE 1.1 — Generic Fruit and Vegetable Canning Flow Chart

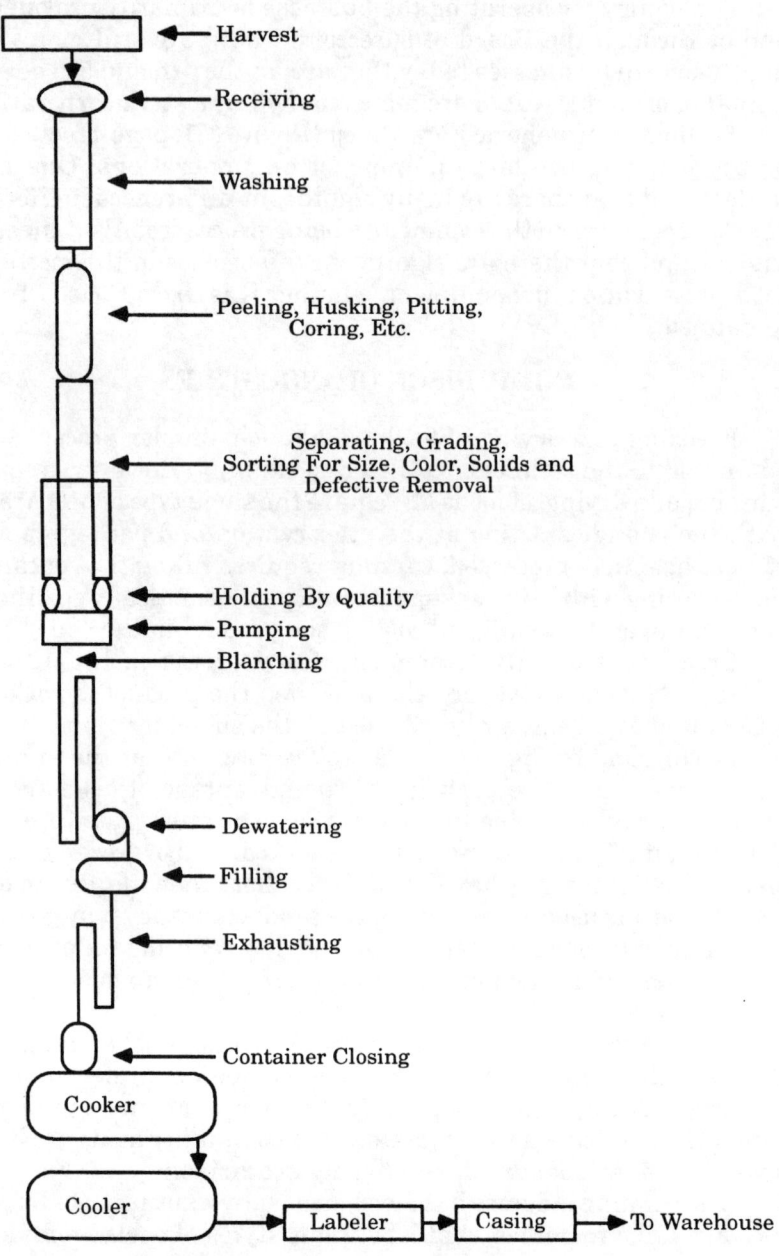

catsup, BBQ sauces, salsas, juices, etc. Today there are well over 100 styles and types of tomato products in the market place. The frozen and dried food industries, also, have developed many mixtures and blends of individual fruits or vegetables and new products by adding sauces and gravies to the vegetables and dinners. The cereal and snack food industries, likewise, have developed many, many products with wonderful flavors which are gaining wide acceptance.

THE REASONS FOR PROCESSING FOOD

There are many reasons for processing foods besides the development of a business with a good Return on the Investment (ROI) for the owners, such as:
1. To prevent spoilage
2. To eliminate waste
3. To preserve quality
4. To preserve the nutritive values in the raw materials
5. To make seasonal fruits and vegetables available year round
6. To put in convenient form for the user
7. To build in "chef" services
8. To develop new products, that is, cucumbers to pickles, tomatoes to catsup, cabbage to kraut, oats to cherrios, corn and potatoes to chips, etc.
9. To safely put foods away for emergencies, and
10. To increase the value of the product.

BASIC FACTORS IN THE LOCATION AND OPERATION OF A COMMERCIAL FOOD PROCESSING OPERATION

The following are some of the basic factors that must be considered in the establishment of a food processing business:
1. **Adequate Water Supply** – The water must be potable and low in mineral salts (calcium, magnesium, sulfur and iron). Upwards of 2,500 gallons of water are required to process a ton of fruits or vegetables.
2. **Sewage Disposal Facility** Wastes from fruit and vegetable processing facilities are high in organic matter, consequently the BOD is high and this must be lowered before discharging even to municipal systems.
3. **Available Raw Materials** – Primary food processing plants are generally located in areas of the production of the individual fruit or vegetable crop. Production implies sufficient

yields to attract growers to want to produce the crop and produce a crop that meets specific quality standards.
4. **Adequate Labor Supply** – This is not the problem that it used to be, as seasonal labor will move from one area to another quite easily and willingly, if adequate work and pay are available. Labor still may be a problem, as the fruit and vegetable industry never has paid the wages of the heavy manufacturing industries.
5. **Sufficient Capital** – This can be a problem, but all big companies were small once and growth is inevitable if a firm is operating efficiently and effectively. General rule of thumb has been that $25,000 is required per 1,000 cases of a given fruit or vegetable per day. This may seem high, but the fruit and vegetable industry is almost always a seasonal industry.
6. **Adequate Markets** – Today, the food firm looks beyond the borders of his own local area and some even think globally. Marketing requires transportation and this may be a major problem, as most foods in the U.S. today move by truck.
7. **Management and Technical Support** – A fruit and vegetable firm must have, at least, the following personnel on board: General Manager, knowledgeable in working with money, people and the given commodity; Superintendent of Production, with ability to work with production problems, that is, manpower, materials, methods and measurement practices; Procurement personnel to deal with agricultural production concerns for each given crop; Quality Assurance and/or Technical personnel

TABLE 1.1 — Organization Chart for A Food Plant

Board of Directors
(Money)

General Manager
(Materials, Methods, Manpower, Machinery)

Procurement	Processing	Quality Assurance	Marketing
Raw Products	Receiving	Line Control	Brokers
Ingredients	Preparation	Product Evaluations	Direct Sales
Containers	Filling	Market Audit	
Labels	Processing	Sanitation	
Machinery	Warehousing	Waste Disposal	
Labor	Labeling	R&D	
	Shipping	Customer Relations	

to deal with regulations, standards, and quality assurance; and either <u>Brokers or In-house Marketing & Sales</u> personnel. Anyone can manufacture a product, but if you cannot market it at a profit, the firm will soon be out of business.
8. **Community and Area Acceptance and Support** – The community and area must be attractive to the firm in terms of infrastructure, that is, roads, electricity, water, sewers, and all other services. In addition police protection, tax structure and tax incentives are most important. Area, State and Local Government must be favorable for an industry to survive.

SOME BASIC REQUIREMENTS OF ALL FOOD PROCESSING EQUIPMENT

The Current Good Manufacturing Practices CFR 110.40 states that "all plant equipment and utensils shall be so designed and of such material and workmanship as to be adequately cleanable, and shall be properly maintained." My interpretation of this statement includes the following:
- All food contact surfaces of equipment and utensils shall be constructed of stainless steel or other materials which are smooth, impervious, non toxic, non corrosive, non-absorbent and durable under normal use conditions.
- Food contact surfaces shall be easily cleanable, and shall be free of breaks, open seams, cracks or similar defects.
- Food contact surfaces shall not impart any odor, color, taste, or adulterating substance to the food.
- Food contact surfaces shall be readily accessible for manual cleaning other than food contact surfaces designed for cleaning in place (CIP) cleaning.
- All joints and fittings shall be of sanitary design and construction.
- In addition, there shall be no dead ends and all food contact surfaces must be protected from any lubricant.

DESCRIPTION OF THE FOOD PROCESSING INDUSTRIES ACCORDING TO BUREAU OF CENSUS STANDARD INDUSTRIAL CLASSIFICATION (SIC) NUMBERS

2032   Canned Specialty Products, such as Baby Foods, Nationality Specialty Foods, Health Foods, and Soups, except Seafood

2033   Canned Fruits and Vegetables, Fruit and Vegetable Juices, Catsup and Similar Tomato Sauces, Preserves, Jams and Jellies
2034   Dehydrated Fruits, Vegetables and Soups (from Dehydrated Ingredients)
2035   Pickles, Sauces and Salad Dressings
2037   Frozen Fruits and Vegetables, including Fruit Juices
2038   Frozen Specialties, that is, Frozen Dinners and Frozen Pizza

TABLE 1.2 — 1992 Census Data for Specific SIC Numbers

SIC No.	Total Number of Establishments	Total Number All Employees	Production Worker Hours (M)	Value Added by Manufacture (B $)	Value of Shipments (B $)
2032	220	20,800	35.5	3,224.1	6,300.3
2033	684	63,900	109.8	6,970.2	14,876.4
2034	150	11,400	18.7	1,133.0	2,334.5
2035	375	21,100	32.2	3,640.5	6,244.3
2037	255	48,000	79.0	2,935.9	7,598.0
2038	362	46,700	73.3	4,100.1	7,838.3
TOTALS	2,046	211,900	348.5	22,004.4	45,191.8

Food processing under any of the SIC numbers normally begins with the procurement of the given fruit, vegetable or other item. Most food processors demand given varieties or cultivars for any given fruit or vegetable because of quality and yield characteristics. The crops are contracted 3 to 6 months or longer prior to processing, based on the needs and demands of the food processing firm. This primary processor prepares, packs and processes the crop directly into consumer units, units for the hotel, restaurant, or institution (HRI) trade, or they may bulk pack the product for secondary processing before or after formulating or engineering the product for the intended purpose.

In recent years, however, both canners and freezers may buy bulk quantities of Individual Quick Frozen (IQF) commodities or concentrated pulps and pastes for secondary processing into dinners, mixed commodities, sauces and catsups, and manufactured into soups etc. Today, much of the juice and many of the drinks in the market are remanufactured from concentrates made at the point of production. These concentrated products, either frozen or canned, are reconstituted at the receiving point and packaged into consumer containers for the ultimate enjoyment of the customer. Thus, the secondary processor eliminates all of the preparation steps. Another advantage is that the crop can be shipped in bulk to the secondary point of processing, saving extra cost in transportation. The secondary processor is the real "value added" processor. In many cases these plants are much more compact, as they are largely a packaging and marketing plant.

MAGNITUDE OF THE INDUSTRY

With the six SIC Nos., there are well over some 30 Unit Operations involved, with many of these the same for each SIC No. and, obviously, several specific to the given SIC No.

There are some 2,049 firms utilizing some 211,900 employees, with the value of shipments over $45,191,800,000.

Further statistical details are found in Table 1.2.

Chapter 2

MATERIALS HANDLING

INTRODUCTION

Brady, writing for The Material Handling Institute, states that "Man has been confronted with the problem of moving himself and materials since the beginning of time. Down through the years he has learned to apply the ancient principles of mechanics — the lever, the wheel, the pulley, and the inclined plane — to make his job of moving, shifting, lifting, and carrying both easier and faster."

DEFINITION

Materials handling is a system of methods and equipment which ties together productive and nonproductive operations and makes them into a processing operation. Brady states, "Material Handling is the art and science involving the movement, packaging and storing of substances in any form" or the organized movement of materials from one place to another. It starts at the source of raw materials, through transportation, receiving, perhaps into and out of storage, movement from one unit operation to another through the food process, into warehouses and out on to the common carrier, into storage waiting distribution and finally out to the customer. It may involve loading, unloading, vertical lifting or lowering, moving horizontally forward and backward, or any combination of the above. Material handling systems integrates the whole processing factory."

OBJECTIVES

Without question, the handling of materials is the single most primary operation of all the unit operations in most food processing plants. The objective of all material handling operations should be to reduce labor costs, material costs, and overhead costs. Further, through the use of efficient material handling systems, there should be an increase in productivity, improved working conditions, increased safety, and, ultimately, an increase in the availability of the product.

Some basic questions must be answered about all material handling operations, such as,
1. Is there any product damage during handling,
2. Is there any accumulation of waste from the material handling operation,
3. Is the material handling operation safe to both the product and the people handling it,
4. Will this system reduce accidents, prevent strain, and improve working conditions,
5. Is the material handling operation efficient, and can it reduce overhead costs and/or,
6. Can this material handling operation be eliminated, combined with another, simplified, or can the sequence of handling be changed to improve the system?

TYPES OF MATERIALS

Food materials are generally of two types, that is, they are either solid or liquid. Solid food materials upon arrival to a food plant may be moved in their containers directly to the processing line for further handling by monorail, hydraulic lifts or they may be unloaded from the receiving vehicle onto belt or screw conveyors, pumped and/or flumed into the factory.

Liquid food materials, obviously, can be pumped, siphoned, or flumed to and through the food plant. Dry homogeneous materials or powders are generally moved by pneumatic transfer systems.

Every commodity has its own peculiarities, and the most efficient systems will greatly depend on the type of material and volume of the commodity to be moved. Materials for processing may be handled in bags, boxes, hampers, crates, barrels, and/or bulk in tankers, railroad cars, or trucks.

RECEIVING

Receiving is the first unit operation in any processing operation. Receiving has two major functions: determination of the level of quality as to its acceptability and secondly, the actual quantity of the load.

MATERIALS HANDLING

The processor needs to know if the load of merchandise is acceptable to the firm, that is, Acceptable Incoming Quality (AIQ). If the load does not meet the firms' requirements, it may be rejected as such or it may be accepted with appropriate deductions. Standards of quality promulgated by the U.S. Department of Agriculture are in existence and many firms rely on a third party to ascertain the quality of the incoming materials. The third party usually is a USDA Receiving Point Inspector. The inspector is looking for quality, absence of defects, size, condition, and/or some internal characteristics, such as sugar content, solids content, etc. If the load is accepted, then it is weighed in and the actual quantity is determined.

The load may be unloaded by washing it out of the vehicle; by lifting the vehicle and rolling the product out (Figure 2.1); by using special equipment, such as the "spudnik," as used for potatoes; by fork-lifting, if in cases, boxes or on pallets; or by unloading by hand. Of course, liquid and powdered products are handled differently, using gravity or pumps.

FIGURE 2.1 — Truck Lift for Unloading
(Courtesy Vanmark Corporation)

14 UNIT OPERATIONS FOR THE FOOD INDUSTRIES

STORAGE BINS AND FEEDERS

A number of solid material feeding operations may be required before moving solids from one position to another in most factories (Figure 2.2). Usually, a large surge capacity is required at one or more points in the process to supply materials on demand to various unit operations. Obviously, for efficient and smooth operations, the solid material should be available on demand for a smooth running operation. The system must be carefully designed to prevent such problems as:
1. No flow or erratic flow due to a stable arch or rathole developing within the solid causing the flow to stop or flow unevenly,
2. Segregation of solids in a hopper or bin with separate fractions feeding out at different times or from different outlets,
3. Degradation, that is, spoilage, caking, oxidation or attrition when materials remain in bin too long. This is usually caused by not following first in, first out sequence (FIFO).

FIGURE 2.2 — Storage Bins (Courtesy Vanmark Corporation)

Johansen indicates that most bins are of the funnel-type, illustrated in Figure 2.3. As can be seen in this figure, the solid flows toward the outlet through a channel that extends upward from the feeder gate. The channel expands upward from the feeder or gate and expands from the outlet in a circular shape and is surrounded by non-flowing material. As the solid flows, the level in the channel drops and layers of the non-flowing material slough off the top of the surrounding mass and slide into the channel. This creates the first in, last out flow. In funnel flow, there is no remixing of segregated material in the hopper. Funnel flow bins can develop ratholes when the non-flowing solid consolidates sufficiently to remain stable after the channel empties. Funnel flow bins are acceptable for coarse, free flowing stable solids that do not segregate.

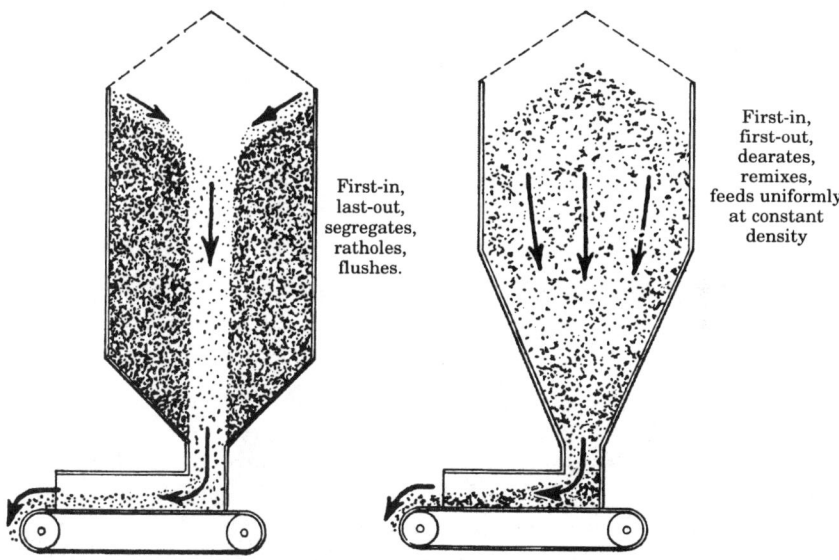

FIGURE 2.3 — Funnel–Flow Bin Design is most common.

FIGURE 2.4 — Mass–Flow Bins do not channel when material discharges.

Johansen describes "a mass flow bin when all the solid is in motion whenever any of it is drawn out" (Figure 2.4). He states that funnel flow bins have a first in, first out flow sequence and they do not channel. They are well suited for powders that tend to fluidize and solids that cake or degrade while in storage.

"Once the bin has been selected, the feeder must be designed to produce flow across the entire bin opening. With small circular or square outlets, almost any feeder will produce uniform flow if a vertical spout about one outlet diameter long connects the bin and feeder. With rectangular bin outlets, however, the feeder must be designed to have increased capacity in the flow direction. A constant

16   UNIT OPERATIONS FOR THE FOOD INDUSTRIES

pitched screw as shown in Figure 2.5 will draw only from the back of the bin. Similarly, the belt feeder in Figure 2.6 will draw solids essentially from the front of the bin. Even with the best designed

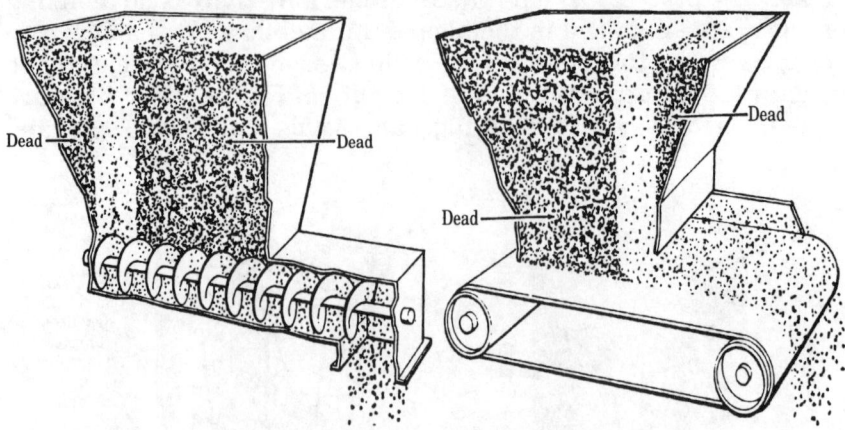

FIGURE 2.5 — Constant Pitch Screw Feeder gives this flow pattern.

FIGURE 2.6 — Belt or Apron Feeder may produce dead spots in hopper.

bin, flow may stop because these localized flow patterns cause stable arches and ratholes to develop in non-flowing regions. Examples of feeders that are designed to cause flow across the entire bin outlet are shown in Figures 2.7 and 2.8. The variable pitch screw feeder (Figure 2.7) shows an increasing pitch in the draw direction. Pitch variation is generally limited to a range between 0.5 dia. minimum and 1.5 dia. maximum.

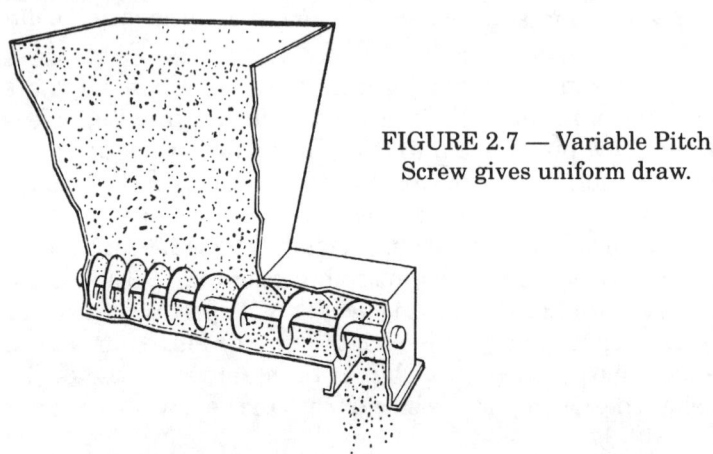

FIGURE 2.7 — Variable Pitch Screw gives uniform draw.

MATERIALS HANDLING

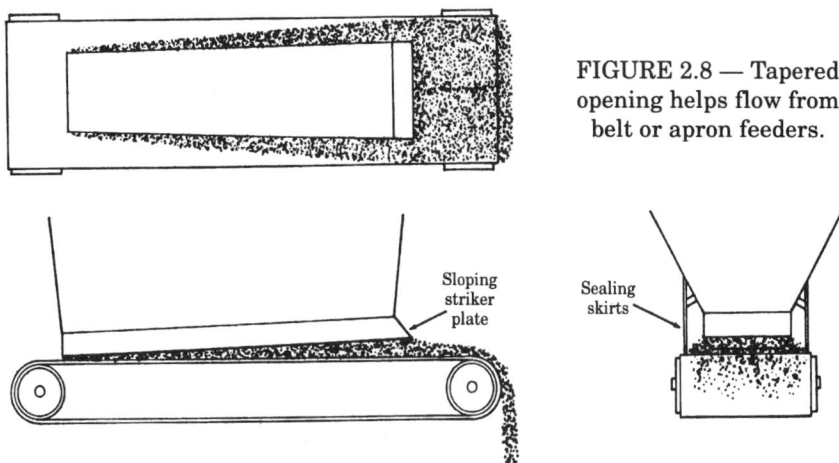

FIGURE 2.8 — Tapered opening helps flow from belt or apron feeders.

"Belt or apron feeders (Figure 2.8) can be given increased capacity by tapering the outlet in horizontal or vertical planes. The sloping striker plate at the front of the hopper is necessary for non-free flowing solids to insure flow along the front bin wall. Taper may be in one direction only.

"Table feeders can utilize the same principles. The skirt is raised above the table in a spiral pattern to provide increased capacity in the direction of the rotation (Figure 2.9). The plow is located outside the bin and plows only the material that flows from under the skirts.

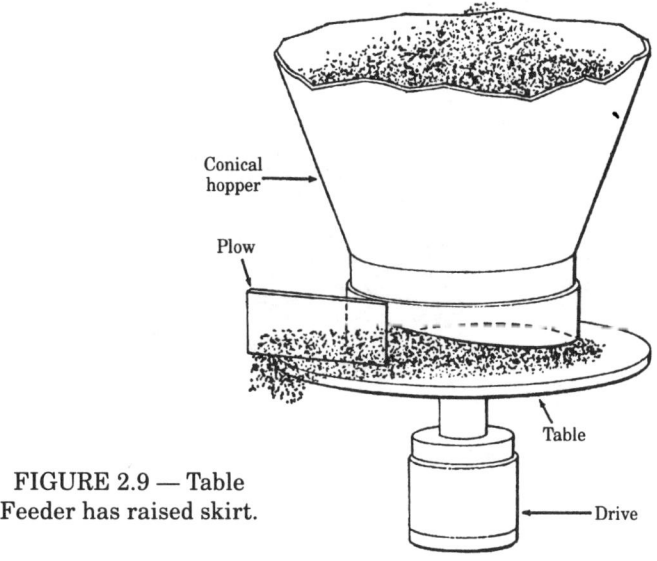

FIGURE 2.9 — Table Feeder has raised skirt.

18 UNIT OPERATIONS FOR THE FOOD INDUSTRIES

"Vibratory feeders can also provide uniform flow along a slot opening of limited length, as shown in Figure 2.10.

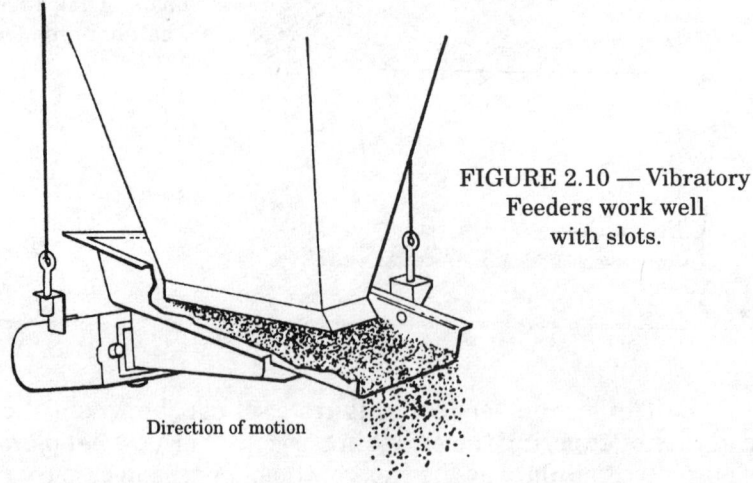

FIGURE 2.10 — Vibratory Feeders work well with slots.

"Star feeders, with a collecting screw conveyor shown in Figure 2.11, provide for very uniform withdrawal along a slot opening. A vertical section of at least one outlet width (and preferably more) should be added above the feeder to insure uniform withdrawal across the opening."

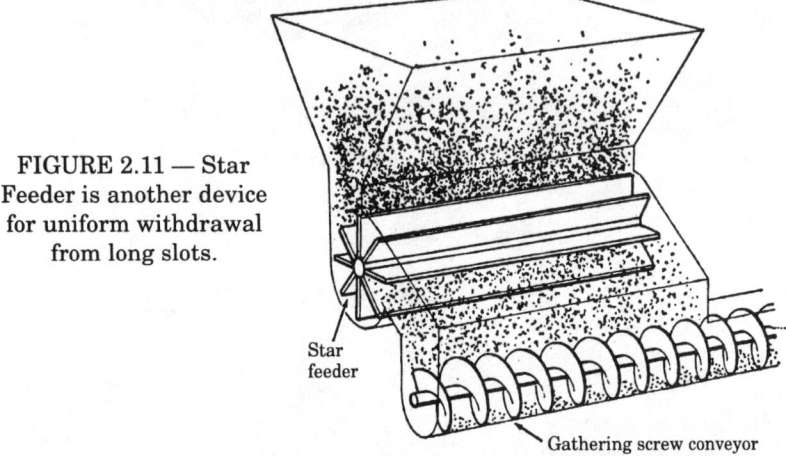

FIGURE 2.11 — Star Feeder is another device for uniform withdrawal from long slots.

MATERIALS HANDLING

SOLID MATERIAL HANDLING

The equipment for the further handling of solid materials may be classified as follows:
- Screw conveyors (Figure 2.12)
- Belt, Chain, Roller conveyors (Figure 2.13)
- Drag conveyors (Figure 2.14)
- Bucket elevators (Figure 2.15)
- Pan conveyor (Figure 2.16)
- Vibrator conveyors (Figure 2.17)
- Flight conveyors (Figure 2.18)
- Water or Flumes (Figure 2.19)
- Monorails
- Hydraulic, i.e., crane, hoist, lift truck, hand trucks, fork lift
- Pneumatic conveyors, which can be classified into three types: (1) Material into air streams induced by vacuum or under positive pressure, (2) Air and material are intermixed simultaneously, and (3) Air into a stored mass of material to cause flow.

Note 1: There are many changes and improvements in conveyors worth noting, such as, High Torque Driven (HTD) belts to replace chains. These new belts have been reported to have a 70:1 life over conventional stainless steel chain driven sprockets. The HTD belts run faster and smoother. Further, they save many dollars in maintenance and energy costs. They need no lubrication and, thus, dirt and grit are totally removed from the system.

Note 2: Another new material handling system that reduces breakage, improves sanitation, and provides greater flexibility to the handling operation are the new vibrator conveyors. They utilize electromagnetic drives and stainless steel troughs as the carrier. The motion of the carrier is created by the elements of frequency (the number of cycles per second) and amplitude (the distance of each cycle's stroke) which works in tamden to balance one another out and, subsequently, the entire system.

[See following pages for Figures 2.12 to 2.19.]

20 UNIT OPERATIONS FOR THE FOOD INDUSTRIES

Standard Pitch, Single Flight

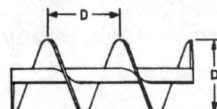

Conveyor screws with pitch equal to screw diameter are considered standard and are suitable for a wide range of materials in conventional applications.

Short Pitch, Single Flight

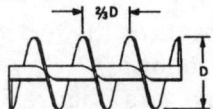

Flight pitch is reduced to 2/3 diameter. Recommended for inclined or vertical applications. Used in feeder screws. Shorter pitch retards flushing of fluid-like materials.

Half Pitch, Single Flight

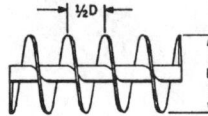

Similar to short pitch, except pitch is reduced to 1/2 standard pitch. Useful for vertical or inclined applications, for screw feeders and for extremely fluid materials.

Long Pitch, Single Flight

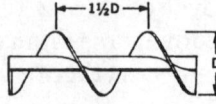

Pitch is equal to 1½ diameters. Useful for agitating liquids or for rapid movement of very free-flowing materials.

Variable Pitch, Single Flight

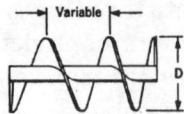

Flights have increasing pitch and are used in feeder screws for uniform withdrawal of fine, free-flowing materials over the full length of the feed opening.

Double Flight, Standard Pitch

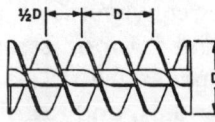

Double flight, standard pitch screws assure smooth, regular material flow and uniform movement of special types of materials.

Tapered, Standard Pitch, Single Flight

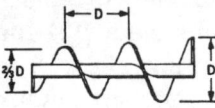

Screw flights increase from 2/3 to full diameter. Used as feeder screws for uniform withdrawal of lumpy materials. Generally equivalent to and more economical than a variable pitch flight.

Single Flight Ribbon

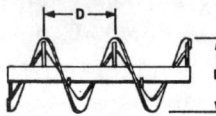

Best for conveying sticky or viscous materials. Open space between flighting and pipe eliminates build-up of the material on the flights.

Cut & Folded Flight, Standard Pitch

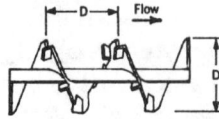

Folded flight segments lift and spill the materials. Partially retarded flow provides thorough mixing action when heating, cooling or aerating light substances.

Single Cut-Flight, Standard Pitch

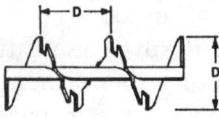

Screws are notched at regular intervals at outer edge for mixing action and agitation of material in transit which tends to pack.

Standard Pitch with Paddles

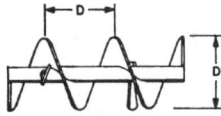

Adjustable paddles positioned between screw flights oppose flow to provide gentle but thorough mixing action. Paddles can be set at any angle to produce desired agitation.

Paddle

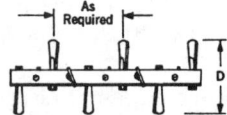

Adjustable paddles provide complete mixing action, and controlled material flow.

FIGURE 2.12 — Basic Conveyor, Flight and Pitch Types

MATERIALS HANDLING 21

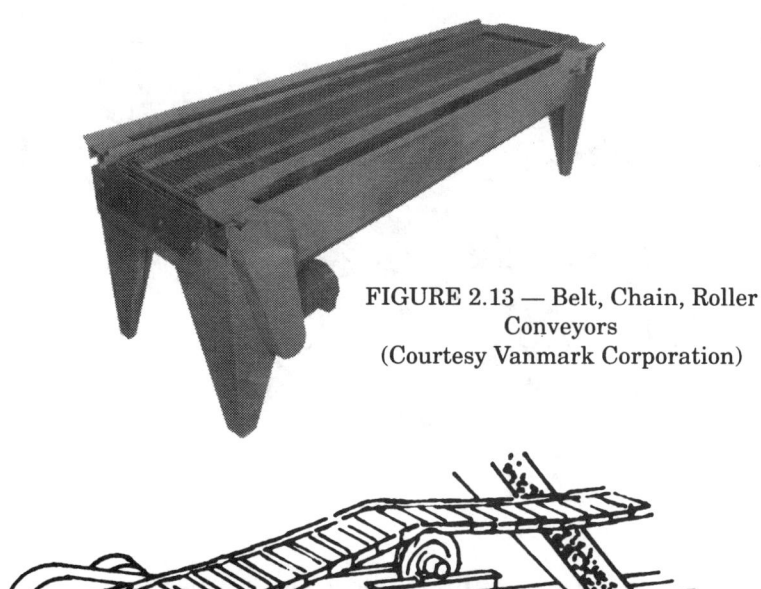

FIGURE 2.13 — Belt, Chain, Roller Conveyors (Courtesy Vanmark Corporation)

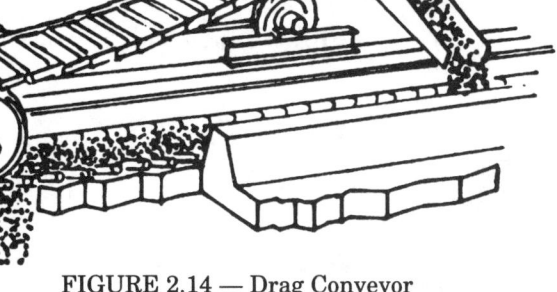

FIGURE 2.14 — Drag Conveyor

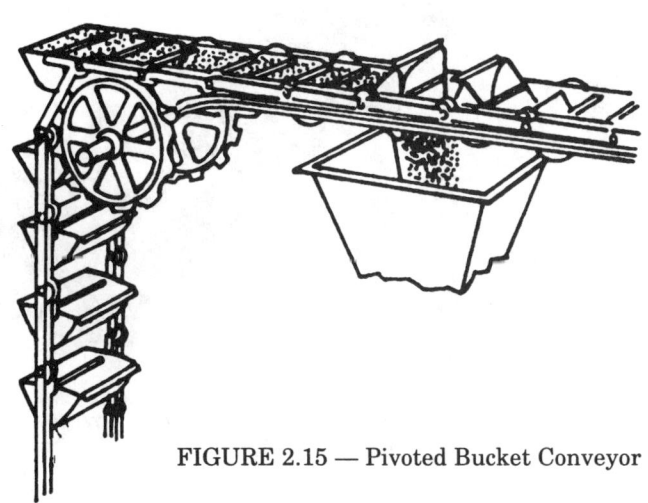

FIGURE 2.15 — Pivoted Bucket Conveyor

FIGURE 2.16 — Pan Conveyor

FIGURE 2.17 — Vibra-Free Distribution Conveyor
(Courtesy AK Robins)

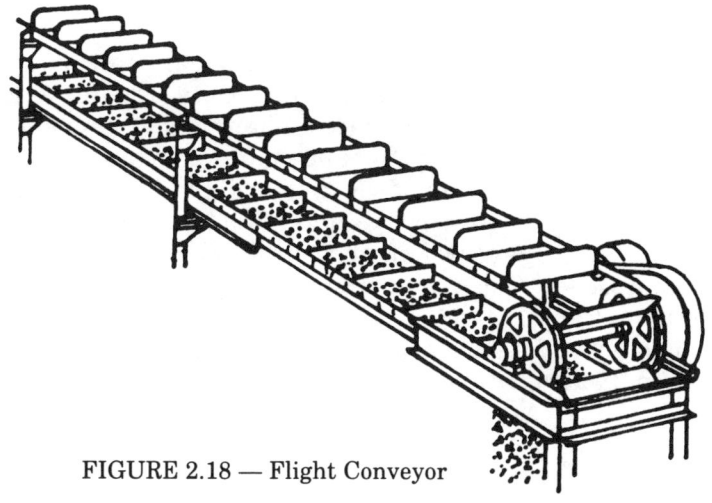

FIGURE 2.18 — Flight Conveyor

FIGURE 2.19 — Water or Flumes

24 UNIT OPERATIONS FOR THE FOOD INDUSTRIES

LIQUID MATERIAL HANDLING

(See chapter on pumps.) Liquid materials can be flumed, pumped, siphoned, or fed by gravity into and through the factory (Figure 2.20). The design of any liquid handling system should be simple, direct, and easy to control and maintain in a sanitary condition at all times. The extent of the system depends in great part on the consistency of the product, volume of liquid being handled, and the amount of storage contemplated or utilized. Knowledge of effects of temperature, corrosion, and viscosity are essential to the proper installation of any system. Wherever possible, the system should utilize safety precautions, including flow regulators, traps, magnets, and foreign material eliminators.

FIGURE 2.20 — Hydro Transfer System Series 7000
(Courtesy Vanmark Corporation)

MATERIALS HANDLING 25

In any material handling system, allowance must be made for storage of the commodity while in process, that is, to take up the slack or over-production at the unit operations before or after, therefore, a surge unit operation is most valuable (see Figure 2.21).

FIGURE 2.21 — Even-Flow Storage and Surge Tank
(Courtesy AK Robins)

Chapter 3

CLEANING

INTRODUCTION

Cleaning is one of the most difficult of all the unit operations in a food plant. It is complicated due to results from the effects of weather conditions at harvest or practices employed during production. Recent emphasis by USDA would indicate that cleaning has more to do with the microbiological problems associated with the given commodity.

DEFINITION

Cleaning is the removal of all harmless extraneous matter and foreign materials from the given commodity.

OBJECTIVES

Food materials are cleaned to remove all "soil", that is, pesticide residues, field soil and dirt, bacteria, molds, yeast, insect eggs and their residues and fragments, and defective tissues. Secondly, commodities are cleaned to improve their appearance and to prepare the product for further processing.

METHODS OF CLEANING

There are two basic methods of cleaning food intended for production, that is, dry cleaning or wet cleaning.

Dry Cleaning

Dry cleaning starts with removal of stones from the product. There are two basic ways to remove stones from a given commodity as shown in Figure 3.1 and Figure 3.2. The equipment in Figure 3.1 is basically a pit or tank where the raw materials are flumed or dumped into. The stones being heavier than the commodity remain at the bottom of the pit while the commodity is lifted vertically and fed on into the next unit operation.

28 UNIT OPERATIONS FOR THE FOOD INDUSTRIES

FIGURE 3.1 — Destoner
(Courtesy Vanmark Corporation)

Dry cleaning may involve separation of the wanted from the unwanted by air under controlled conditions (Figure 3.2). The air may be blown across the product, it may be blown through the product as it moves through the factory on belts, or it may be blown through the product as it falls from one level to another.

Air separation is an economical method of removing much of the unwanted material from given commodities as the lighter particles may be blown out by a blast of air while the heavier pieces will fall through the blast of air. An excellent example of this system is the cleaning of blueberries when hand harvesting them. Using the picking rack (similar to a dust pan with fingers on it), the leaves are separated from the berries after sweeping the rack through the blueberry bush by emptying them from the rack when holding the rack high and letting the berries fall into a container with air (wind) blowing through them as they fall. The leaves are blown away and the berries are collected in the container for further handling.

CLEANING

FIGURE 3.2 — Air Blast Cleaner
(Courtesy G. J. Olney, Inc.)

Some dry cleaning may be done at the time of harvest by the producer. The amount of cleaning can be directly related to the quality of the seed used at harvest. Seed used for planting given crops should not contain weed seeds.

Dry cleaning is, also, affected by moisture on the commodity at the time of cleaning. The drier the product the easier it is to dry clean. As would be expected, dry cleaning is, also, greatly influenced by the amount of commodity being cleaned and the amount or velocity of air blowing over or through the product. Air velocity must be adjusted for each commodity depending on the specific gravity and particle size involved.

Dry cleaning is used to clean empty cans and jars prior to filling by blowing or jetting air into the inverted empty container. It is only effective as long as the foreign matter is loose and removable by air.

Wet Cleaning

Wet cleaning as the name implies means wetting the fruit or vegetable in an effort to solubilize the "soil" to aid in removing it from the given commodity. "Soil" may mean spray residues, insect or rotted areas, mud, sand, dirt, and harmless extraneous matter, that is, leaves, trash etc. This soil must be loosened and then washed away or separated from the given commodity.

A system that combines removal of trash that will float followed by stone and mud ball removal is shown in Figure 3.3. A vibratory screen follows the stone and trash removal to remove water from the commodity. Obviously, the riffles must be watched carefully and the stones removed as needed to make this system work. This system is found in many plants because of its good efficiency and through-put.

FIGURE 3.3 — Model DW-150 with Stone Crib (Courtesy G. J. Olney, Inc.)

Wet cleaning requires water that is potable and free of bacteria and impurities, both organic and inorganic.

Detergents, acids, and alkalies are most effective adjuncts or aides in loosening the soil. Specific detergents may be formulated for given "soil" removal by each commodity. Further, elevating the temperature of the water and application of appropriate agitation will greatly enhance the "soil" removal.

In actual practice there are generally three steps in cleaning or the removal of "soil" from most food materials. First, the product should be "wetted" or soaked for given periods of time, preferably in a solution containing a specific detergent that will loosen "soil" on the commodity. Secondly, the temperature of this flume water may be elevated by injecting live steam directly into the flume water. A better method may be to inject the stream as to agitate the commodity in the flume or even in the wash water. Compressed air may be used instead of steam to accomplish the agitation to help remove the "soil" but the temperature effect is lost. This is one of the reasons many food processors use flumes in their factories as they are soaking the product as they move it into the factory.

Third, the soil must be washed away. This is usually done with high pressure water sprays while the given commodity is conveyed out of the flume water. There are three fundamentals that effect the removal or the efficiency of the washing operation. First, the distance of the nozzle head from the commodity should be such as to effectively cover a given area for every type of nozzle(s).

My experience has shown that an effective nozzle should not cover more than one square foot of surface area at the point of impingement. The size of the particle size, and the amount of pressure impacting on the commodity are the other two variables that effect the efficiency of the washing operation. Particle size should be as small as possible, but large enough to loosen the soil. The water droplet should not penetrate the commodity. The amount of pressure should be great enough to loosen the soil and aid in carrying it away.

For example, when washing tomatoes the following suggestions are offered to free the tomato of all soil. Water temperature in the flume at 110–115 °F (43–46°C), with a dwell time of 2 minutes in an alkali detergent of 5%, followed by a spray wash using a GG18 Square Fulljet nozzle having a capacity of 6.4 GPM at 150 psi. The GG18 nozzle is manufactured by Spray Systems in Chicago IL.

Another method of wet cleaning involves flotation wherein the undesirable can be floated away. Sometimes detergents (high foamers) are helpful to remove thistle and nightshade berries. Wood can be floated away from root crops prior to further processing. The water should be soft and sufficient dwell time should be given for greatest efficiency.

As implied above, empty containers should be cleaned before filling and they can be cleaned much more effectively by using detergents to first loosen the "soil" followed by high pressure washing. The "soil" in food containers is a result of handling practices from manufacture to filling. Practices in use should be all aimed at keeping the empty containers clean at all times.

Chapter 4

QUALITY SEPARATION

INTRODUCTION

Fruits and vegetables and many other food commodities differ widely in quality characteristics. These differences are due in great part to the variety or cultivar. Climate, area of production, production practices, maturity, and/or stress conditions during growth and maturation also may have a direct effect on the quality of the commodity.

DEFINITION

Quality separation is a unit operation designed to separate distinctive quality characteristics or attributes of a product. These may include color, size and/or length, absence of defects, and/or internal composition, that is, sugar, solids, acidity, etc.

OBJECTIVE

The whole objective of quality separation is for the manufacturer to produce a product that is uniform, as uniformity of quality is most important when packing most products. No customer wants large and small, immature and mature, or good and bad in the same container or package. The industry strives for uniformity through sorting and grading for the many differences that may exist in the raw materials.

METHODS

Generally, the biggest single quality difference is size and length of the individual unit within a commodity. Secondly, the difference may be due to maturity, wherein color, texture and specific gravity may be ways of classifying the differences. Thirdly, differences due to defects and defective areas within a unit may require the need for quality separation practices.

34 UNIT OPERATIONS FOR THE FOOD INDUSTRIES

Modern technology has allowed the food processing industry to critically distinguish differences within the commodity using mechanical/physical equipment, electronic means of sorting, and chemical separators. Many of these separators are simple and very effective in distinguishing differences within the commodity.

Generally, quality separation can be classified into the following major areas:
1. Size
2. Color
3. Defects
4. Specific Gravity
5. Weight

Each quality separator system is quite specific in how quality may be separated. Sorting for quality has moved from the producer gate to the factory because it's more efficient, accurate, and economic. The following are some examples.

Size Separation

Size separation is usually accomplished with holes in drums or belts, screens, or cups. Size is a quality characteristic that may indicate wide differences in quality. For example, a size 1 pea is a pea that will pass through a $9/_{32}$ inch (7.1 mm) hole, while a size 7 pea

FIGURE 4.1 — Rotary Size Grader
(Courtesy AK Robins)

QUALITY SEPARATION 35

is larger than a $^{14}/_{32}$ inch (11.1 mm) hole. There can be seven sizes of peas in any given load. Each of these sizes may be further separated into one or more specific gravity lots to separate out maturity within a given size. I have seen as many as 28 lots of peas packed from a given load, based on seven sizes and four specific gravity separations within each sieve size. (See Figure 4.1)

Size separation is usually the first quality separation of most round commodities, that is, apples, apricots, dry beans, lima beans, beets, berries, cherries, figs, grapefruit, mushrooms, peaches, peas, pears, peppers, pickles, pineapple, plums, potatoes, olives, onions, oranges, strawberries, and tomatoes. This size separation may be by simple screening (Figure 4.2), or by use of diverging belts or rollers.

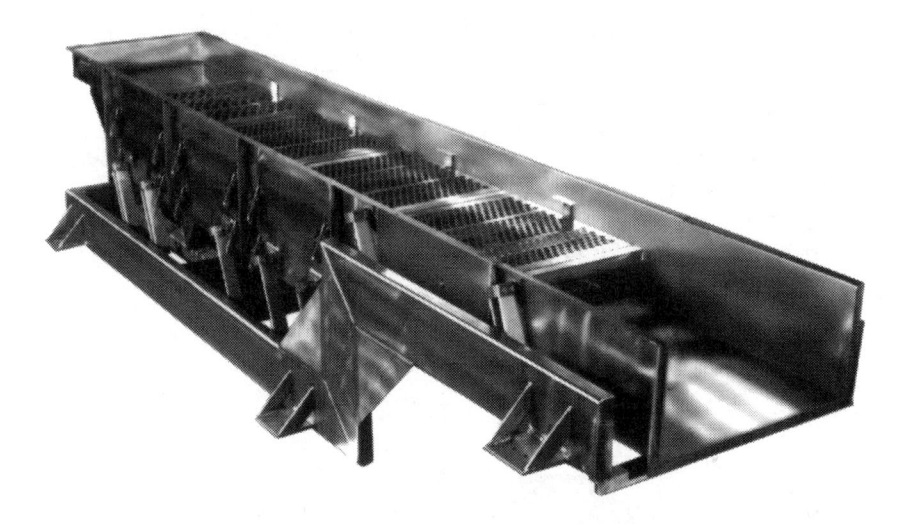

FIGURE 4.2 — Vibra-Free Two-Deck Grader
(Courtesy AK Robins)

Some commodities, like asparagus, green or wax beans, carrots, and cucumbers, are separated for length in addition to size. Length is, also, an indication of quality, with the shorter lengths usually designated higher in quality.

Screening and mechanical sorting for size and length is, without question, the most economical system available to the food processor. However, they are quite specific in what can be accomplished.

Vision Separation

Many commodities are separated by vision, using color sorting machines for color and defects. Specific colors usually designate maturity or ripeness. Defects may be due to stress, mechanical handling, or diseases and insects, and generally show up as brown or light brown discolored areas. The defects on other products may have different colors, such as, greening on potatoes.

The new electronic vision sorters (Figure 4.3) have eliminated the human touch in color and defect sorting and the sorters can be most specific in color or defect separation. Electronic sorters are speedier and much more accurate. However, they may be expensive for the seasonal operator, but, with the labor problems prevalent in some areas of the country, sight, electronic or vision sorters may be the answer to the future to separate quality differences, at least, as far as color and defects are concerned. It should be pointed out that these types of sorters can do much more than separate color and defects, as they can separate out foreign materials and even internal structures that are defective. The electronic sorters remove the defective or off-colored product by air jets or vacuum suction. Electronic sorters are just now coming in use in many of the food processing plants. They are as effective as the operator wants them to be.

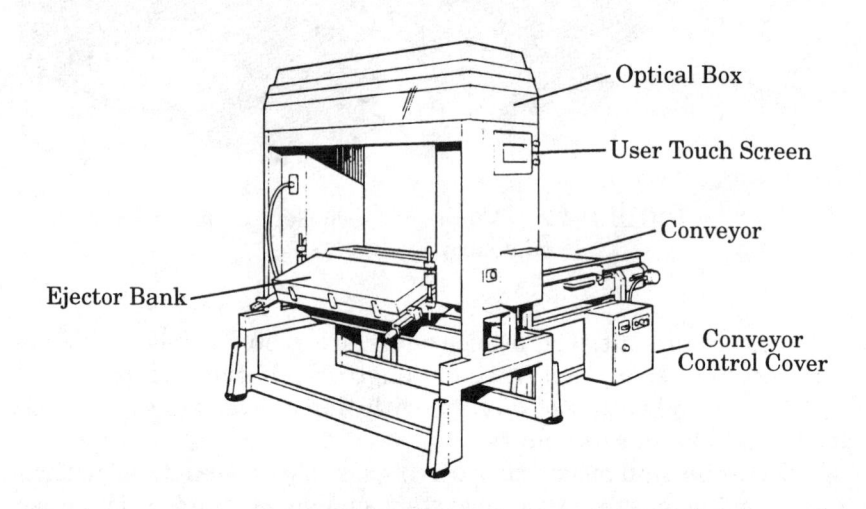

FIGURE 4.3 — Sortex 6000 Electronic Vision Sorter
(Courtesy AK Robins)

Specific Gravity Separation

Specific gravity separation has been in use for many years, using brines to separate the immature from the mature in products like peas and lima beans. It is based on the principle that the heavier products are more mature and will settle in given solutions. Specific gravity differences are significant with potatoes within a cultivar or among the various potato producing areas. Processors use the specific gravity technique to sort loads for uniformity in French Fry and potato chip manufacture. The specific gravity technique is very useful when purchasing potatoes for soup manufacture or other processing operations where differences in solids is important.

Weight

Weighing is one technique to separate qualities utilized for some fruits and vegetables and eggs. Weighing may be done with the individual unit or by package, or even case lots. Some equipment is available, but weighing is not one of the more widely used separation techniques in commercial food processing.

Most quality separations are completed prior to the product being blanched or before further processing. Exceptions exist, such as, specific gravity after blanching, as the occluded air must be removed for accurate separation. Another exception is defect and color removal of chips after they are manufactured. Incidentally, the latter is a poor practice, as much energy has been wasted in frying the defective raw potato slices, the oil may be contaminated, and/or through-put is reduced in direct proportion to the amount of defective or throw-away product.

Uniformity in quality is the key to satisfying the customer. Separating by size, length, color, absence of defects or solids content is the first step in assuring quality of most commodities.

Sorting and Hand Trimming

Sorting and trimming are the removal of the undesirable portion(s) of the product from the desirable portion(s). As indicated above, there are many machines designed to do the sorting and trimming; however, much of it is still done today by people (inspectors, graders, trimmers and/or sorters).

Sorting by eye-balling is a major means of separating and practically always used, regardless of the other more sophisticated methods. It is most important that the product move slowly and that the product should be rolling through the use of rollers (ideally

38 UNIT OPERATIONS FOR THE FOOD INDUSTRIES

these rollers should be color coated to match what is desired, that is, the color of the given product so that the worker has a constant standard in front of them at all times) with all sides of the product observable. The speed of the belts may be increased from 20 ft./min. as the particle size goes up (peas and beans), to 25 to 30 ft./min. for tomatoes and oranges. (See Figure 4.4)

FIGURE 4.4 — USDA Dumping and Continuous Grading Belt for Inspection and Grading

Lighting over the inspection belts is an important key to workers' efficiency and it should be relatively higher in intensity for older workers, that is, 75 foot candles to as much as 125 foot candles. Most importantly, the light source should accentuate the color differences between the defects and the normal color, rather than mask the inspector's or sorter's eyes, according to the USDA. The color temperature of the light is important, as low intensity (3,000°K) brings out the brown colors while intensity at 6,500°K accentuates the green colors. USDA Galleon Brown states that dark colored produce requires twice as much light intensity as light colored produce.

Further, the worker should be adequately trained in what to sort or trim out, if any, of the commodity. Such things as types of defects, misshapen items, off colored, over or under matured, etc. should be used to train the worker and they should be instructed in exact tolerances, if any. The use of videos and models are most worthwhile. Lastly, it should be noted that hand sorting and trimming are tedious jobs and everything possible to make the worker comfortable, with well planned relief, will make for greater efficiency.

Chapter 5

PEELING

INTRODUCTION

Peeling is a unit operation that is common to many fruits and vegetables. The peel or the outer protective covering of a given fruit or vegetable protects the given commodity from injury during handling and storage. Many commodities have peel that may be inedible and thus it needs to be removed before the product can be consumed.

DEFINITION

The peel is usually removed as it may be inedible or it is removed to make the product more attractive to the user. In the case of most root crops the peel contains much "soil" and possible microorganisms that may be harmful to mankind. Peeling may be defined as removal of the inedible portion from the edible portion.

OBJECTIVES

The objectives of peeling are to remove all of the inedible portion and as little of the edible portion as possible. Peeling is a costly unit operation and the removed peel may be difficult to dispose of or find a use for.

The following commodities are usually peeled prior to preservation: apples, apricots, beets, carrots, corn for hominy, grapefruit, peaches, pears, peppers and pimentos, pineapple, Irish potatoes, sweet potatoes, and tomatoes. The methods used to peel these commodities may be quite different. Further, the amount of peel removed varies from very little to all peel removed.

METHODS

There are six basic methods in use today for peel removal: (1) hand knife, pliers or peeling spoons; (2) mechanical using a knife to follow the contours of the given commodity; (3) brush using brushes having varying degrees of stiffness; (4) carborundum abrasion or grinders to grind off the peel; (5) chemicals using alkalis and acids to digest components of the peel and the pectic substances just below the layer of peel; and (6) steam, atmospheric or high pressure to cook or blow off the peel.

No system is perfect and the amount of loss of usable food may be the difference between profit or loss for a processor. Peel loss within a given commodity is directly associated with the size of the item being peeled. The small sized items within a given commodity, due to surface area, may lose 4 or 5-fold the loss from the larger item within the same category. Secondly, the shape of the commodity, that is, the roughness or contours of a given item may make it almost impossible to peel profitably with some of the above methods.

The amount of the peel removed with the knife, or batch, or continuous peeler, great care should be exercised in controlling the proper adjustment or the depth of the cut or abrasion so as to only remove the peel or pericarp. This is most important for efficient recoveries. Peeling too deep removes valuable edible tissues.

Brushing and abrasion peel removal is a common method for peeling many root crops (see Figures 5.1 and 5.2). The success with these peelers is the proper use of the brushes or abrasion rollers.

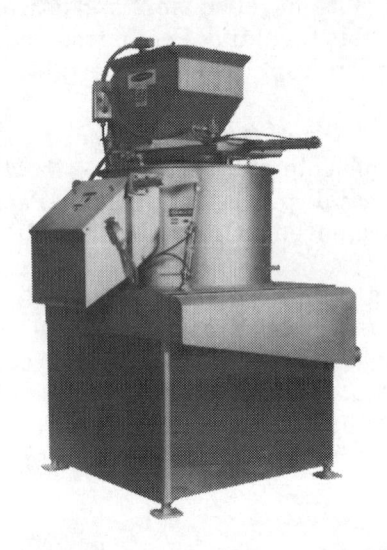

FIGURE 5.1 — Batch Peeler (Courtesy Heat & Control)

PEELING

FIGURE 5.2 — Continuous Brush and/or Abrasion Peeler
(Courtesy Vanmark Corporation)

Soft nylon brushes are used on freshly harvested crops while hard brushes are used on stored crops. Likewise, light carboundum stones are used on the freshly harvested crops while heavy carboundum stones are used on stored commodities. Another important facet of peeling with brushes or abrasives is the dwell time in the peelers. Longer dwell times will eventually grind away much of the usable commodity. The RPM of these peelers is most important to recoveries. Speeds above 250 RPM may be detrimental in terms of peel loss. Another important point when using brush or abrasion peelers is the load going through the peelers. These peelers are designed to operate at full capacity and anything less has a tendency to be inefficient and may ultimately damage the product. Obviously, excess peel removal creates a problem of waste disposal, to say nothing of the poor recoveries.

Chemical peeling (Figure 5.3) is an old technique, but finding much wider use for many crops because of the many variables the user has as options to improve efficiencies, that is, concentration of the chemical (alkali or acid), temperature of the solution, and dwell time in the solution. The operator should take advantage of this information, but always remember that if you change dwell time, concentration or temperature you will probably have to change one of the other variables for efficient peeling. Over-exposure to lye may result in a poorly peeled product due to oxidation or loss of product.

FIGURE 5.3 — Lye Peeler/Scrubber/Washer for peeling tomatoes, other fruits and root crops. (Courtesy Leader Engineering & Fabrication Co.)

Wide differences exist, even in the application of the chemicals as to the commodity with some chemical peelers utilizing complete immersion while others only spray the chemical on the commodity. In some cases even wetting agents are used to assure complete coverage (wetting) of the chemical on the commodity. Efficiencies of chemical peelers are much superior to any of the above systems with considerably less waste, shorter time, and less costly. However, chemical peelers do present a problem in regards the disposal of the peels and/or the chemical solution. Another concern with chemical peeling of fruits and vegetables is the safety factor. These chemicals are corrosive and reactive and great precautions must be exercised in regards the equipment and the personnel working around and with these chemicals. The lowest concentration possible should always be used from a safety standpoint. Following chemical peeling the product is generally washed. In some cases it may be necessary to neutralize the chemical during or following the washing by giving the commodity an acid wash if it was alkali peeled or vise versa. One other important point to remember when using chemicals for peeling is that the chemical generally will be used up and the concentration must be checked and corrected periodically to obtain uniform results.

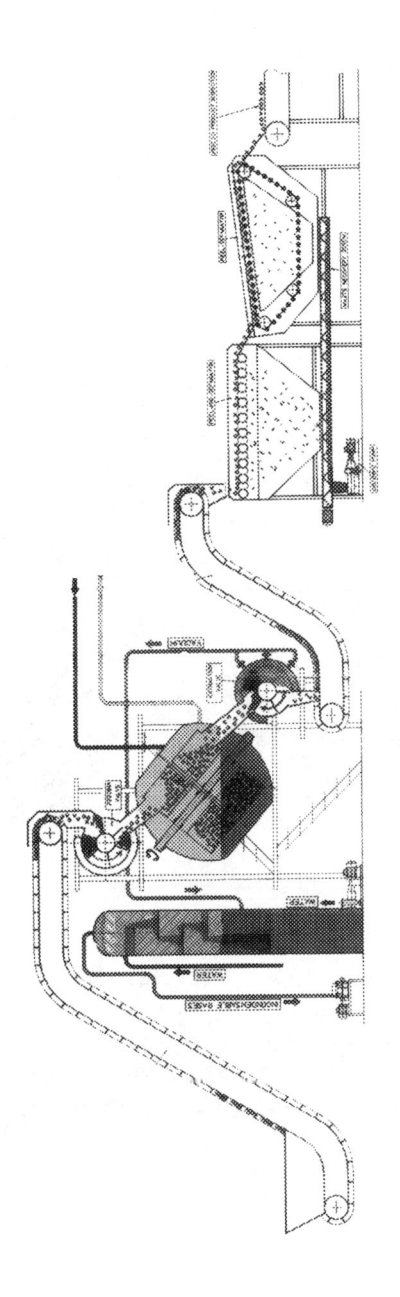

FIGURE 5.4 — 'Saturno' Thermal Peeler (Courtesy Savi Antonio, srl)

46 UNIT OPERATIONS FOR THE FOOD INDUSTRIES

Steam peeling has been widely used over the years. With the introduction of high pressure steam peelers (see Figure 5.4, previous page, and Figure 5.5), they are very effective for peel removal and the finished product may be superior to other methods in terms efficient peel removal. Further, there is no chemical residue in or on the commodity and the amount of peel removal can be very carefully controlled. The problem with the high pressure peelers is that they are batch designed and do require more time in loading and unloading than the other systems.

FIGURE 5.5 — Vacuum Peeling Line (Courtesy IMBEC)

In either chemical or steam peelers where heat is applied, the commodity must be cooled quickly to prevent the heat from penetrating the flesh and creating other problems down the line in terms of soft flesh causing poor qualities, recoveries or yields. As stated above neutralization of the acid or alkali may be a must.

TABLE 5.1 — Relationship of Tuber Size to Various
Physical Constants for Tubers of Different Diameters

Physical Constants	Sphere Diameter of Potatoes in Inches					
	$1\frac{1}{2}$	2	$2\frac{1}{2}$	3	$3\frac{1}{2}$	4
Surface Area (sq. in.)	7.06	12.6	20.4	28.1	39.2	50.0
Volume (cu. in.)	1.76	4.19	9.16	14.13	23.82	33.51
Ratio of Area to Volume	4.00	3.00	2.50	2.00	1.75	1.50
Approx. No. Tubers/Lb.	11.00	8.00	5.03	2.60	1.80	1.00
Approx. No. Slices/Lb.	264	256	201	125	101	64
Approx. No. Tubers/8 lbs.	88	64	40	21	14	8
$\frac{1}{16}$" Peel Removal (% Volume Loss)	21	17	14.5	12	10.5	9

Chapter 6

DISINTEGRATION WITH LITTLE CHANGE IN FORM

INTRODUCTION

There are a whole lot of unit operations in a food processing plant that directly affect the processing of many commodities. These unit operations are operations necessary to prepare the product for preservation.

DEFINITION

Disintegration means to separate into component parts or break up the individual unit. In this chapter we are talking about disintegrating by separating with little change in form. They include such operations as husking of corn; pitting of cherries, prunes and other fruits; coring of apples, peaches and pears; shelling of peas and lima beans; snipping of snap beans; and de-stemming of cherries, berries, cocktail grapes, and olives. There may be other unit operations with little change in form, such as inspection (hand, mechanical and electronic), dirt eliminators, and oriented feeding systems. Most of these unit operations are specific and used for given commodities. Each is necessary for preparing the product for further processing.

OBJECTIVES

The objective is to separate the commodity without damaging the main edible component. There are basic requirements for each of these separations or unit operations, such as:
1. The unit operation must not damage the finished quality of the product.
2. The unit operation must be consistent and effective in producing uniform units of the given commodity.
3. The unit operation must be economical and labor saving.

4. The unit operation must be designed to conform to modern sanitary design and the unit must be easy to keep clean and in working condition.
5. Most importantly, the unit operation must be suitably priced and affordable to the industry.

TYPES OR METHODS

Husking

This unit operation should be completed in the field, however, it is still accomplished at the factory because of need for preservation of product quality. About one-third, depending on cultivar and maturity, of the harvested corn is husk and is of little or no value within the factory. As a matter of fact, all husks are hauled back to the field or farm for feeding or silage. Husking at the factory is a costly phase of a corn canning or freezing operation.

Nevertheless, corn must be husked and the new machines are very efficient and much less damaging to the kernels than earlier models. Generally, the ears are oriented to feed tassel end first into the husker and the modern rolls on the husker grip the flag leaves and strip the corn of its husk. Most corn husking machines will cut the butt end off the cob and, if properly adjusted, they are most effective in removing only the butt end of the cob and not kernels. The operation of a corn husker is much improved if the machine is kept slightly wet.

Immediately after husking, due in part to the bruising of the kernels, the corn should be further processed to preserve quality. If it is to be cut, the sooner it is cut and blanched, the greater the retention of product quality. If it is to be packed, canned or frozen as cob corn, it should be thoroughly blanched to arrest any further degradation of the sugars and to inactivate the enzymes.

In my opinion, there is no better product on the market than sweet corn (whole kernel, cream style, or cob corn), but it must be properly processed. This includes the correct variety or cultivar, harvesting at peak maturity in terms of sugar development, husking quickly, and followed by complete inactivation of enzymes to preserve what nature provided. Much of the corn today could be significantly improved by following basic know-how and it all still starts with coordinating the husking operation with the harvest and further processing.

Shelling

Shelling has shifted from the factory to the field with the greatly improved pea and bean combines. These combines are hydraulically powered and operated under wide terrain conditions most efficiently. This change in practice from factory shelling to field shelling has greatly improved sanitation and cleanliness in and around the factory and has left the vines and pods in the field where they belong. These combines are magnificent machines and they are truly the work of great engineering.

Peas and lima beans are harvested at peak maturity by the combines moving through the fields while mowing and thrashing the product in one unit operation. The peas or beans are collected within the machine and eventually they are emptied and transported to the factory. The secret of success is the time from thrashing to the factory. Quality retention is temperature related and it is most essential that this be preserved by rapid handling practices.

FIGURE 6.1— Pea Harvester (Courtesy Klockner Hamachek)

The thrashing operation consists of a series of beaters to break open the pods. Obviously, when beating a pod, some damage naturally occurs to the individual pea or bean. This damage consists principally of triggering enzyme activity to generate chemical changes that may affect flavor and loss of sugars. If extensive damage exists and the time is lengthy, even detrimental odor development may occur.

Modern harvesting operations cope with the new practices quite well and today overall quality for peas and beans is the best ever. Another reason for the great success today in retaining the quality of both peas and beans is that most of the production is in the northern tier of states, particularly the Northwest and Minnesota and Wisconsin. Here, cool nights generally prevail and the harvesting is in early Summer. Cool temperature not only preserves quality, but lessons bruising of the product during harvesting, and greater recovery should be assured at the factory level.

Pitting

Pitting is common to cherries, peaches, apricots and plums for canning or for dried prunes. Each machine in the market place is specifically designed for the given commodity and in many cases the machines are only leased.

The machines are quite simply designed, requiring the fruits to be lined up properly or centered in pockets or holes and the pit or stone is punched out of the fruit. It is a fast operation with little damage to the fruit.

In the case of peaches and other fruits, the pit or stone may be removed by halving the commodity and cutting or shaking the pit or stone out. Obviously, the product must be ripe and the stone must be free. When the pit or stone is cut out, care must be exercised to remove all the pit or stone. Tolerances exist for pieces, but that should not be the goal when pitting or removing the stone from the product.

With cherries, the secret of success comes by carefully sizing the cherries and removing the deformed cherries. The U.S. Standards for Grades of Red Tart Pitted Cherries tolerates not more than one missed pit for every 20 ounces of pitted product.

Coring

This is a common unit operation for cabbage, apples, pears and some other fruits and vegetables.

The equipment ranges from a simple bit, like when boring a hole in wood or metal, to machines precisely designed to holding the fruit while reaming out the center. In every case, the core is directly related to the variety/cultivar and the size of the individual

DISINTEGRATION/LITTLE CHANGE IN FORM

unit within the commodity. Therefore, for greater efficiency, the commodity should be size graded before coring.

Core removal is a costly operation and the amount removed has a direct bearing on the yields. However, cores are generally too fibrous to consume and in some cases they are not highly edible.

One other consideration should be noted and that is small sized products require the same time and motion as large sizes. Further, the losses are far greater and, of course, the efficiency is severely lowered.

Snipping and Destemming

Stems are natures way of attaching the edible commodity to the plant. All growth of the edible portion comes from the actual leaves and passes through the connective tissues, the stem, to the storage area, the edible commodity. In some plants, the stem is part of the harvested commodity and is generally not eaten, as it is woody and very fibrous and non-digestible. Therefore, it should always be removed from the commodity before consumption.

FIGURE 6.2 — Snipper and Unsnipped Bean Remover (UBR) for Snap Beans (Courtesy G. J. Olney)

With green and wax beans, the stem is easily removed by snipping it off with a machine designed to allow the end to stick through holes and be cut as the beans move through the machine. With small beans, if the snipper is not adjusted for their size, they may all be cut up or snipped in part. Therefore, it is important to properly adjust the snipper for efficient removal. Ideally, only the stem end should be removed, as the blossom end is the growing point and highly edible and digestible. However, snippers are not that good in orienting the bean to allow only snipping of the stem end. Also, it should be noted that all bean clusters should be broken up before entering the snipper for efficient operation. It is almost impossible to have perfectly snipped green or wax beans. Therefore, the U.S. Standards for Grades of Green and Wax Beans sets forth tolerances by Style and Type within each of the Grades. The processor of this commodity should always strive to be well below these tolerances for given label specifications.

FIGURE 6.3 — Stemout Machine Being Used On Tomatoes
(Courtesy Magnuson Corporation)

At one time, stems were a major problem with tomatoes. Today, through genetics, the tomato breeder has found the gene responsible for holding the stem onto the fruit. At the time of harvest, the tomato pedicel (stem) should be jointless and the tomato breaks clear from the vine during the harvesting operation. Thus, fruits are not

DISINTEGRATION/LITTLE CHANGE IN FORM 55

punctured by the stems. Before this knowledge of tomato genes, the stems were removed by a machine similar to the ones still in use for many other fruits. One firm calls their machine, the "Stem-out". It is a unit operation of much value to the olive industry, to the plum and prune industry, to the cherry and grape industry, and to many small fruits for processing.

"The Stem-out employs a bed of small counter-rotating rubber rolls that are at right angles to the direction of product flow. A quick-return, reciprocating motion to the roll bed progresses and gently agitates the fruit, permitting the stems to be gripped between the rolls and removed. The same reciprocating motion that progresses the fruit also drives the rolls, with alternate rolls being gear driven in opposite directions from the enclosed stationary gear racks along each side of the machine. The Stem-out is quickly and easily adjusted for product flow by tilting the bed. After stemming, the product is delivered to a hopper or belt for subsequent operations".

In the early days during the development of this unit operation, I had the opportunity to evaluate this equipment on strawberries and blueberries. We were able to obtain nearly 100% efficiency in the removal of the stems in several plants as we moved over the Eastern part of the U.S. With continued improvements over the years, it's easy to understand the success with this unit operation and the great value of its use for the production of de-stemmed products.

Chapter 7

DISINTEGRATION WITH CONSIDERABLE CHANGE IN FORM

INTRODUCTION

Many unit operations in a food processing operation may make considerable change in form to the product. These may be classified as styles of pack. With the wide variations in food products, the lack of labor, the development of new technology and appropriate equipment, many "new" products are coming to the forefront. In some cases, it is the same old product in a different form as nothing much is added except perhaps a blend of two or more ingredients.

DEFINITION

This unit operation is similar to the previous definition except there is considerable change in form. These can be divided into the following separate parts: (1) Cutting, (2) Shredding, (3) Crushing and/or comminuting, (4) Homogenizing, (5) Sheeting, and (6) Extracting and/or Juicing.

OBJECTIVES

Again the objectives are very similar to those previously described except by considerable change in form, we change the style of the product into what the market accepts as a new commodity. The objective in this unit operation is to make new products from the same commodity. Cream style corn is different than whole kernel corn, etc.

TYPES OR METHODS

Cutting

This is an essential unit operation for many commodities simply because the product may be too large for a given package, or it is cut into a more convenient form for the user, or it cannot be efficiently processed as a whole unit, or new styles of products are in demand.

FIGURE 7.1 — 8-Lane Sizer/Halver (Courtesy Magnuson)

Many examples can be used, but perhaps the best is the massive trend in the vegetable salad business by the producer or preprocessor cutting the commodity at point of production or shipment and making it ready for direct use prior upon arrival. These salad products are washed after cutting and heavily chlorinated prior to packaging. The shelf life is relatively short, but perhaps much longer than the whole product because of the washing and chlorinating. Regardless, they are in demand and many users (hotels, restaurants,

DISINTEGRATION/CONSIDERABLE CHANGE

and institution-HRI) buyers prefer them because of labor savings and readiness for serving. Retail stores prefer this product for their customers as it eliminates many hours of work and much waste over the whole product(s).

Cutting may be in the form of dices and cubes, i e., 1/8 x 1/8 x 1/8 or all the way up to 1 inch x 1 inch x 1 inch, slices both flat and crinkle or wavy types for many root crops, strips for root crops and some salad crops like celery and broccoli, and actual shredding as with many of the salad crops and cabbage.

Many food products are cut, but using different types of cutters due to chemistry of the given product. As an example, some products must be moisturized prior to cutting due to their dryness and the ultimate over heating of the product. Other products may have too much sugar and as they are cut they become very sticky. While other products may be too fibrous and the fiber accumulates in the cutter creating inefficiency and poor cuts.

Without question the Urschel food cutting machines are the most unique line of equipment for this unit operation.

FIGURE 7.2 — Model CC Slicer (Courtesy Urschel Labs)

60 UNIT OPERATIONS FOR THE FOOD INDUSTRIES

There are over 40 models and each can cut given products in wide ranges. There are many other manufacturers for specific cutting operations, but overall the wide Urschel line of equipment is worth investigation. The equipment can be completely made of stainless steel and usually is for use in given food operations. The different models cut food to very strict tolerances such as the CC Model for potato slicing at 0.060 inch for chip frying with variations less than 0.004 on mature and high solid potatoes. Differences greater than this may create wet centers or slices that will not fry properly. This model will slice high quality potatoes at 8,000 pounds or more per hour.

Shredding

Shredding may be accomplished on similar machines as used for cutting, but the product may be torn apart at the same time as it is being cut. Shredded products are common for salads and some fermented items.

Crushed, Chopped or Comminuted

These styles of products are becoming more popular as mechanization of the harvesting operation increases. It should also be pointed out that crushed, chopped or comminuted products have come into being due, in part, to labor shortages.

FIGURE 7.3 — Tomato Chopper (Courtesy FMC - FPMD)

Personally, I think it is a far better utilization of crops, particularly fruits, that are fully ripened and too soft for whole or diced or sliced packs. They generally have excellent color, flavor, and they are ideal for cooking purposes, such as, soups, pies, side dishes, and puddings.

Crushing of products by chopping them into small pieces (0.025 to 0.075 inch) prior to heating, speeds enzyme activation. Therefore, direct super steam introduction into the chopper or crusher may aid in more thoroughly inactivating the enzyme before the enzyme breaks the product down creating poorer quality products. At the time of crushing and immediately after comminution, removal of the air is most important to prevent oxidation and other chemical reactions within the tissues. Some have claimed that super heated steam aids in creating a partial vacuum. Obviously, super heated steam, preferably under a vacuum, will eliminate the air and aid immensely in creating a better consistency in the final product.

Homogenization

This unit operation is an out growth of the dairy industry and has found many useful purposes for a more complete break-up of the particles in some liquid items like tomato juice and other liquid products. Homogenization is the process of putting incompatible or immiscible components into a stabilized suspension in a liquid medium. Homogenization depends in great part on the pressures or resistance's and orifice openings in the homogenizer. Homogenizers generally function by passing the product under pressure between closely clearing but relatively fixed surfaces. The high velocity, hydraulic shear, pressure release, and impact, rend the dispersed phase into a very fine state of subdivision of the order of 1 micron in diameter (about 1/25,000 in.) Most of the homogenizers function by pumping the fluid mixture under high pressure to a narrow opening between a valve plug and its seat, the size of the opening being controllable. Upon examination of homogenized products, many cells are completely broken apart; however, the products are superior in that there is little or no separation and, of course, no large particles or cells in the finished product.

Emulsification

Emulsification is that process of size reduction in which two or more miscible liquids are intimately mixed, one is the dispersed or discontinuous phase, the other being the dispersing or continuous medium. In most emulsions one of the immiscible phases is usually aqueous in nature. Milk is an example of an oil-in-water type of emulsion, while mayonnaise and unworked butter are examples of water-in-oil emulsions. The particle size of the dispersed phase of most common emulsions is of the order of 1 to 10 microns in diameter. The design of equipment for the production of emulsions emphasizes the principle of subjecting the liquid mixture to a vigorous shearing action rather than to the impacting action often stressed in many processes for size reduction of solids. A colloid mill is often used and it works on the principle shearing the liquid product by passing the fluid between two surfaces that move at high velocity and at close clearance with respect to each other, see Figure 7.4.

FIGURE 7.4 — Juice and Pulp Extractor (Courtesy FENCO)

Extraction or Juicing

Juice extraction or juicing, Figure 7.4, may be accomplished with a screw type extractor where the product is forced against a screen of specific diameter as it moves through the unit or the more conventional paddle revolving inside a drum with a given size screen wherein the product is forced against and through the screen. Recently, a belt press extraction system has been introduced to the industry. This system takes the chopped or comminuted product and squeezes the juice from the pulp. Various levels of efficiencies can be expected depending on the pressure applied. See Figure 7.5. Some prefer one type or the other.

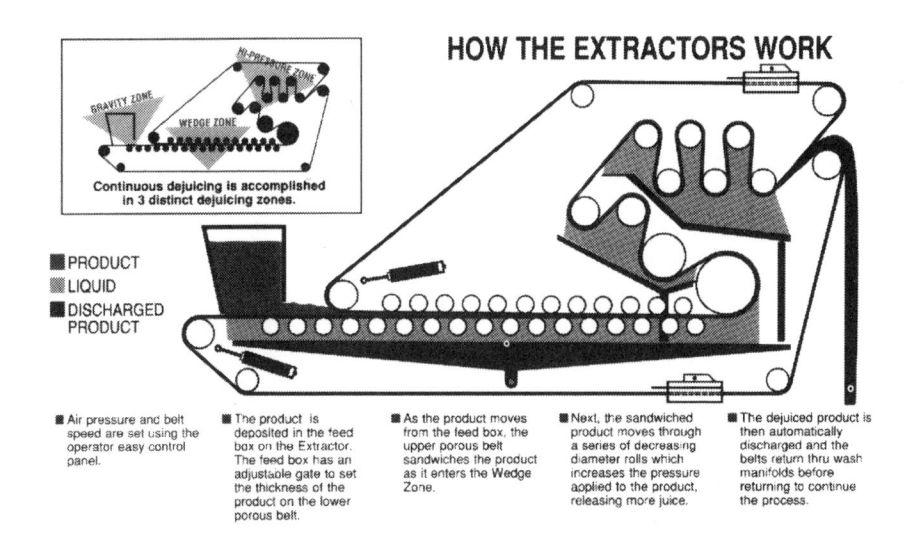

FIGURE 7.5 — Belt Type Continuous Juice Extractor
(Courtesy Frontier Technology, Inc.)

With the paddle type of extraction the paddle moves the product forward in the extractor and forces it through a screen of given diameter. For juice, the screen may be from 0.020 to 0.300 in diameter. The product is held in the extractor by the aid of a gate to build up given pressures inside the extractor. Yields obviously will vary depending on (1) the screen size, (2) the pressure build-up in the extractor due to the gate restriction and/or the paddle speed, (3) the maturity or softness of the fruit due to ripeness and/or preheating of the fruit, and (4) the particle size entering the extractor. Yields also are dependent on the different commodities, their seed size, amount of pericarp or skin, and fruit size. Smaller fruits yield less than larger fruits. Generally, yields will vary from 70 to 95% of juice removed from the original product. In some cases if the extractors are set too high, seeds and skin or peel are forced through or broken up and forced through the screen resulting in a poor quality product with unsightly particles.

Sheeting

This is a unit operation that is used with many commodities to make the prepared product into thin layers or sheets for further processing. Some commodities after crushed or pulped are mixed with water and other ingredients and then sheeted into thin layers followed by cutting into shapes and frying or drying or baking. The sheeting operation may consist of two drums rotating in opposite directions pulling the product through and into preadjusted sheets or thin layers. Following the sheeting of the product, it is cut into triangles, squares, diamonds, circles, etc. for further processing. Corn chips, tortilla chips, and many types of extruded products can be manufactured following sheeting.

Chapter 8

SEPARATION

INTRODUCTION

Most foods are composed of many chemical substances. In some cases, their chemical composition can be predicted by using physical methods. This unit operation finds use for many different types of foods. It could be a part of the unit operation, Quality Separation, but I have chosen to keep it separate because it deals more with chemical composition rather than strictly physical changes.

DEFINITION

Separation means to divide, take apart or break up into its various components.

OBJECTIVES

To separate a food into its various components or parts, such as, solids from liquids.

METHODS

Filtering and Filtration

Proper filtration and clarifying is one of the most important requirements when manufacturing many food items. Filtration is a common procedure in the milk industry, in the fruit juice industry, in the frying industry, etc. Filtration is simply a unit operation of separating the solids from the liquids or portions thereof. Filtration may be an adsorption process, whereby particles are attached chemically to the filter. In our discussion, filtration is a process of physical separation by sieving, clarifying, straining, sedimentation, interception or Brownian diffusion.

The physical separation is simply a case of the particles being larger than the pore size of the medium. In some cases, filtration takes place by gravity flow, in others the product is forced through the filter or screen, and still in others the separation may take place by centrifugation. The basic separation techniques are classified according to the physical phenomena they utilize, that is, from gravity to effective pressure systems.

Clarification is used when solids are to be separated from liquids. Enzymes and/or other filter aides may be added to speed up the separation of the liquid from the solid particles and allowing the product to settle for given periods of time before decanting or further clarification or separation. Temperature is an important aspect of proper separation, as hot liquids separate easier than cold liquids.

Filtration is commonly used on a continuous base when frying foods to remove the loose food particles from the oil. The amount of food particles removed is based on the pore size of the filter. Small micron filtering is most important to retain the quality of the oil in a satisfactory condition. As with any filtering operation, the filter must be changed periodically to be effective. Generally, these continuous filters are pressure operated and the changing of the filters is based on the amount of head pressure required.

Deaeration

Deaeration is the removal of occluded and dissolved air from food products prior to packaging the food into containers. Deaeration is distinguished from exhausting in that, during exhausting, the air removal is primarily in the head space, while in deaeration, air is removed from within the product before filling. Deaeration is limited to liquid and particulate type foods. Deaeration prevents browning of foods due to oxidation or contact of the food with oxygen.

Visible oxygen in foods affects the appearance of the product, particularly if it is packed in glass.

Oxygen in foods may create foaming of the food prior to filling resulting in uneven filled containers. The solubility of air in foods decreases as the temperature increases and at the boiling point all the air should be removed from the food. With carbonated beverages, deaeration of the water helps the product retain more of the carbon dioxide.

The Foxboro Company published a report showing some of the reasons for deaerating several food items as follows:

Orange Juice–Canned, Whole	Elimination of foaming
	Less scorching during pasteurization
Milk and Cream	Removal of off-flavors and off-odors
Catsup	Eliminates separation
	Prevents "black neck" or oxidation in the neck of the bottle
Tomato Juice	Better retention of Vitamin C (Ascorbic acid)
	Eliminates foaming at filler
Baby Foods	Better color
Apple Juice	Retards corrosive action
Chocolate Syrup, Honey, French Dressing	Eliminates unsightly air bubbles
Table Syrup, Pineapple and Grape Juice, Peach and Apricot Nectar	Eliminates ring of foam in neck of container

Deaerators consist of a vacuum chamber into which the product is sprayed or dispersed. Entrapped air easily escapes as the product is flashed off or vaporized in the vacuum chamber. While flashing off any off-flavors, off-odors and air are released and these can be removed through the vacuum outlet, while the deaerated product settles to the bottom of the deaerator for removal.

Gas packing involves air removal by evacuation and replacement by an inert gas, such as nitrogen, immediately before sealing. This is commonly used for heat-sensitive, dry or some solid pack items.

Deaeration by vacuuming is replacing some exhausting equipment. Pre-vacuumizing is used in the packing of some fruits. After the air has been removed from the fruit and the vacuum broken, syrup enters the space formally occupied by the air and superior products are the result. This process is referred to as vacuum syrups and is used for fruit products today.

Chapter 9

IN-LINE PROTECTIVE EQUIPMENT

INTRODUCTION

This chapter is all about manufacturing safe food. Safe food is essential to satisfy consumers and produce products that are free of potential health problems which are referred to as extraneous materials or potential contaminants. The Current GMP now states that "effective measures shall be taken to protect against the inclusion of metal or other extraneous material in food. Compliance with this requirement may be accomplished by using sieves, traps, magnets, electronic metal detectors, or other suitable means."

DEFINITION

In-line protective equipment are unit operations that should be installed in production lines to eliminate extraneous materials from contaminating food, either from entering the plant with the raw materials, or from contamination of the products during manufacture.

OBJECTIVES

To prevent a food product from becoming contaminated with metal or other extraneous materials while being manufactured.

METHODS

The following are some suggested separate unit operations that may be helpful to prevent foods from becoming contaminated with metals or extraneous materials and to comply with this regulation.

Sieves and Screens

Sieves and screens are used to separate some food materials into their various sizes and to remove unwanted harmless extraneous matter, stones, and pieces of equipment from the food.

In sieve analysis work, according to Perry, the apertures are understood to be round in a sieve and square in a screen. Dry sieving is used, unless the material is so fine that flocculation interferes. In such cases, wet sieving is employed. A combination of the two procedures is often advantageous; wet sieving to remove materials able to pass through a No. 325 sieve, for example, and dry sieving on the dried retained portions. (See Table 9.1)

Screening may be accomplished using flat vibrating screens stacked up with the larger screen on top, or long vibratory beds with the smaller openings at the beginning of the bed. The screens have various holes, depending on the commodity and sizes desired to be separated. The products may also be separated using revolving reels with screens of various sizes. In this case, the smaller holes at the beginning of the reel with the size openings increasing toward the end of the reel, thus, the larger size usually is discharged over the end. Sometimes, the size graders are nested with the product entering the inside of the nested graders, with the largest product removed first, second largest on the next screen, etc.

Screening and sizing are important attributes of quality, with the smallest size considered the highest quality for all vegetables. Fruits are generally just the opposite, that is, the largest size is considered by the buyer as the best quality. I do not necessarily agree with the latter, as sometimes the large sizes lack sweetness and flavor.

Sifters and Entoleters

This is used in the flour, salt, seasoning, etc. as a means of separating the different sizes and the chaff or other foreign material from the product. Sifting is often considered a cleaning operation.

According to Parker, "high speed gyratory sifters have been designed to remove grain contaminates together with larvae and insect debris from grain. The gyro-type sifter has the additional advantage of not breaking the grain. In practice, if the sifter is operated ahead of the first break rolls, insect-free grain at the rolls is assured." An additional advantage accruing from the use of the gyro-type sifter is its ability to remove the spores of black mold. Grain enters the inlet at the top of the sifter and passes over a series of such sized mesh openings, so as to hold the grain but permit extraneous matter to pass through.

Screenings passing through the wire cloth go to an insect killing unit, commercially called the "Entoleter." This method utilizes centrifugal force, with the machine consisting of a rotor formed by two circular horizontal steel plates. The lower plate is rigidly attached to the hubs fitted to the motor shaft, and the upper plate with open center, is fastened to and separated from the lower plate by two concentric groups of round, solid, hardened steel spacers, called impacters. These impacters are spaced apart around the outer and inner peripheries of the rotor. Grain or flour is fed into the center in a thin film by means of a distributor in the inlet and is then thrown by centrifugal force through the apertures formed by the two plates and the impacters. The treated product is discharged in a uniform spray around the rotor and then carried off through the outlet spout. The particular value of the Entoleter is its ability to kill all forms of insect life.

Magnetic and Metal Detectors

Magnetic and electronic metal detectors are becoming more important in a food factory as more machinery is used in the

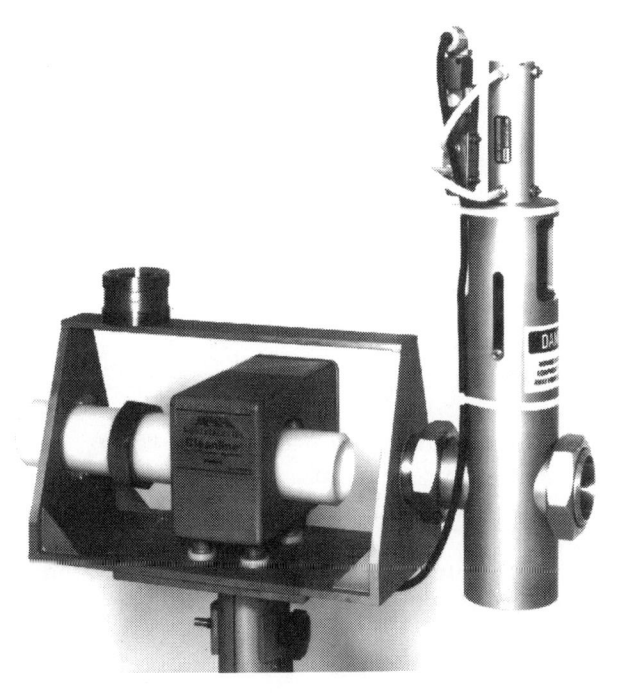

FIGURE 9.1 — Cleanline Metal Tracker System
(Courtesy Barkley & Dexter)

72 UNIT OPERATIONS FOR THE FOOD INDUSTRIES

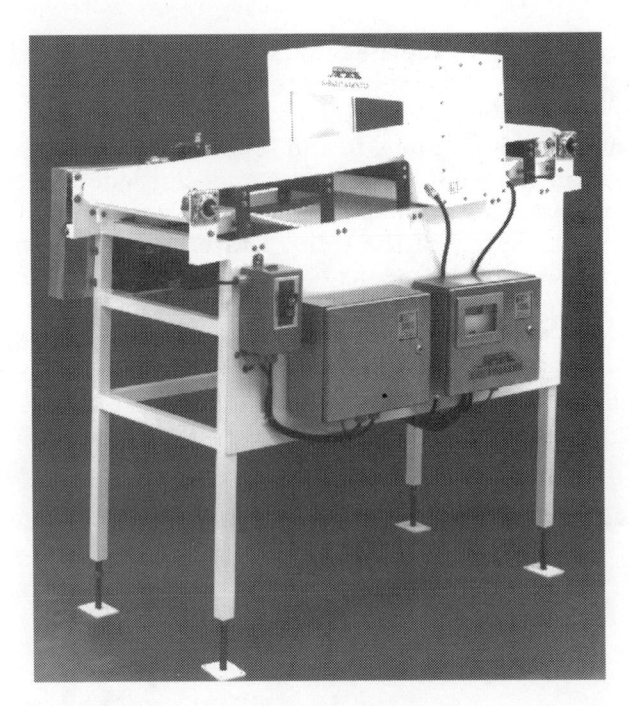

FIGURE 9.2 — Metal Tracker Pass-Through System
(Courtesy Barkley & Dexter)

production, harvesting and handling of the given commodity. Machines may have lose nuts, broken pieces of metal falling off, parts of belts fraying, and even tramp metal, aluminum cans, and pieces of foil and plastic bags from the fields at time of harvest.

Electronic metal detectors use a high frequency electromagnetic field and metals passing through this field disturb it, that is, the metal creates a circuit imbalance and this triggers a reject system, sets off an alarm, or other appropriate signal. The advantages of this system include its ability to: (a) detect both ferrous and non-ferrous metals; (b) distinguish between product and actual metal debris; (c) detect metal in varying size ranges, depending on the aperture or pass-through in the systems search head; and (d) integrate with a product reject component ideally suited to the processing or production methods in use.

IN-LINE PROTECTIVE EQUIPMENT 73

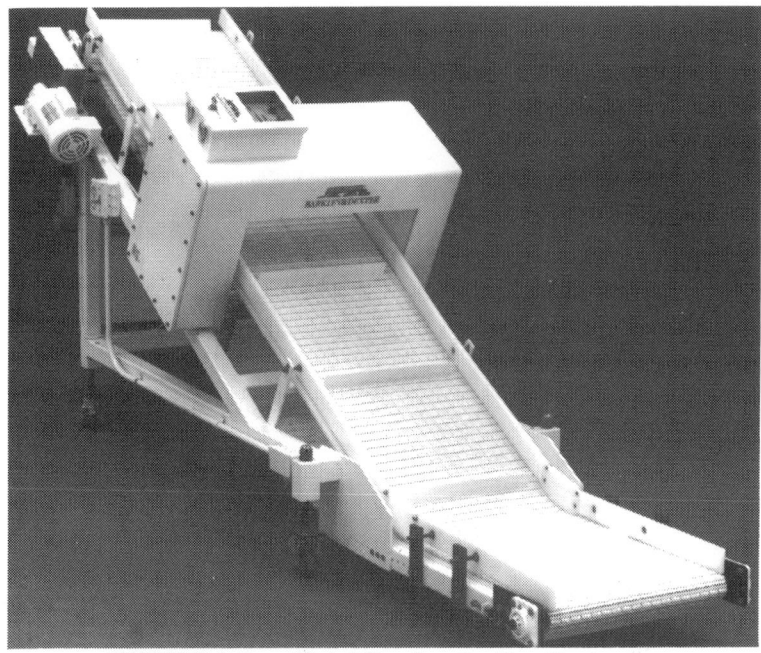

FIGURE 9.3 — Metal Tracker Incline System
(Courtesy Barkley & Dexter)

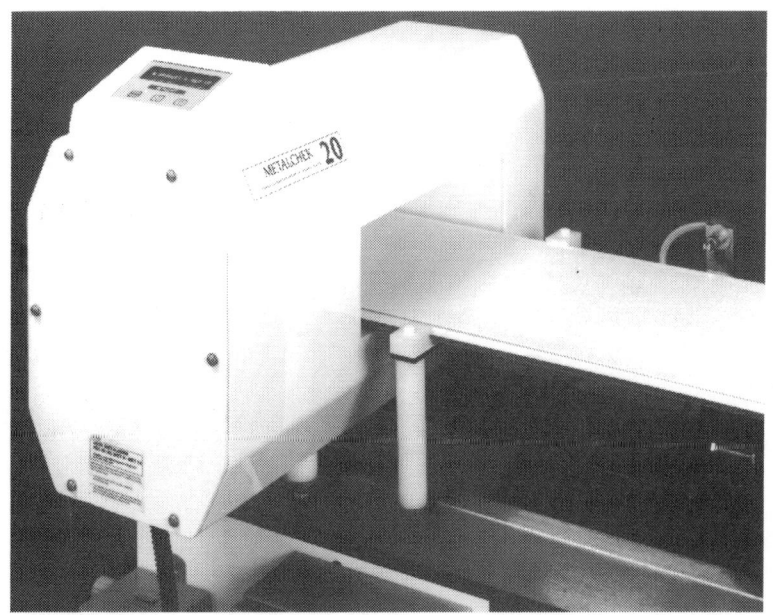

FIGURE 9.4 — Metalchek 20 (Courtesy Lock International)

Chapter 10

BLANCHING AND PRECOOKING

INTRODUCTION

Blanching and precooking are short heat treatments required to inactivate the natural present enzymes and to remove the occluded air that may be present in food products. The word blanching may be synonymous to scalding, steaming, precooking and, in some cases, peeling.

All living matter is made up of various chemical substances. Further, all living matter have enzymes present. Enzymes are protein in nature and they have colloidal properties. They act as organic catalysts and these enzymes are capable of causing many changes in food materials that may be detrimental to assuring their quality. Thus, enzymes should be inactivated as quickly as possible. Inactivation is generally accomplished by the application of heat. However, some enzymes may be inactivated by the use of chemicals and/or the exclusion of air.

There are many classes of enzymes, based on how they react in the particular substrate or food. Food processors should be primarily concerned with the oxidase, pectase, and proteolytic type of enzymes. The oxidase enzyme oxidizes food and destroys vitamin C and causes off-colors and flavors, the pectase enzyme hydrolyzes pectin that affects the texture of foods, while the proteolytic enzyme decomposes protein.

DEFINITION

As used in this unit operation, blanching is a heat treatment or chemical treatment to inactivate or arrest enzymes from attacking a food to cause it to discolor, become changed in texture, or lose flavor.

OBJECTIVES

The first objective of blanching is to inactivate the enzyme. It is impossible to kill the enzyme, as it is a chemical, but it can be made inactive by the heat treatment. Blanching, also, removes gases from the tissues and aids in maintaining vacuum in canned foods. Blanching softens some foods like asparagus and decreases the volume in foods like spinach and, thus, permitting proper packaging. Blanching decreases the microbiological load and thus aids in cleaning the product and removing the earthy, viney or grass-like flavors.

Blanching may be used for many fruits and vegetables for canning. However, the blanch is not necessarily for enzyme inactivation, as the enzyme would be inactivated during the canning process. It is done to remove the off-flavors, expel the occluded air, set the color, improve the texture, and cleanse the product.

Blanching, however, is mandatory for all vegetables and fruits for freezing and drying to preserve their inherent quality.

METHODS

Blanching may be accomplished with hot water (170°F (77°C)) or steam at 240°F (115.6°C). Some products should be blanched in water to preserve color and texture (spinach and green beans are examples), and to remove the occluded gases. Steam blanching is quicker and there should be much less leaching of soluble nutrients from the product.

Some products are given a steam heat treatment, as in blanching, but really all that is accomplished is wilting of the product as used for pumpkin and squash. Other products are blanched to aid in peeling, like potatoes and beets. Most canned fruits are not blanched, although apples and peaches may be blanched to prevent discoloration due to oxidase enzymes. Also, the enzymes in some fruits for freezing may be inactivated by use of sulfur dioxide treatment.

There are many designs of hot water blanchers with the blanching accomplished in either a batch type dip or moving the product through hot water. These latter blanchers are of several types, that is, two belts with the product in between, a pipe or tube with the product pumped though the hot water, or a perforated drum with a screw conveyor moving the product through the hot water while agitating the product. All of these water blanchers have the flexibility of temperature control, and with variable drives on the equipment, the dwell time can be accurately controlled to assure proper inactivation of enzymes.

Steam blanchers have come to the forefront since World War II,

BLANCHING AND PRECOOKING 77

as they are simpler to control for assuring adequate blanch. They may be a simple tunnel with a belt carrying the product through steam, they may be the loading of trays and placing the trays in an enclosed chamber with steam admitted as needed, or they may be an enclosed trough with a screw conveyor moving the product forward while direct steam is admitted to inactivate the enzyme. The advantage of the latter two is that steam can be under pressure.

FIGURE 10.1 — Rotary Hot Water Blancher/Cooker
(Courtesy AK Robins)

All blanchers are energy intensive, with steam blanchers requiring 75 to 90% less energy than water blanchers. Normally, 2 to 8 pounds of steam are required per pound of product, depending, of course, on the particle size and the degree of blanching when using steam blanchers.

Another advantage of the steam blancher over the water blancher is that there is considerably less leaching of solids and nutrients from the product, and thus less of a BOD problem in the effluent.

Most blanched products are immediately cooled following blanching to better preserve the product quality. Cooling is accomplished by spraying cold water over the product, dipping or immersing the product in cold water, or pumping the product through cold water. Frozen and dried products should be thoroughly cooled to 50°F (10°C) or lower for better retention of quality prior to freezing or drying.

Chapter 11

PUMPS AND PUMPING

INTRODUCTION

Pumps and pumping represent one of the most important aspects of nearly all food processing operations. Some consider pumps and pumping as the heart of many food operations. Certainly, when all the pumps are added up in terms of number of unit operations performed or in terms of horse power, this unit operation is most significant. Proper pumps and pumping on schedule makes for a smooth-running operation.

DEFINITION

Pumps are a mechanical system of moving liquid and semi-liquid products in and around a food plant. Pumps are of many types and have many uses. All things considered, pumps are almost mandatory in the modern food plant to handle many products.

OBJECTIVE

The objective of pumping is to move liquid and semi-liquid products from one unit operation to another, or to move the product through a given unit operation such as a homogenizer.

TYPES OR METHODS

There are many types of pumps used in a food plant and there are many uses for pumps. Figure 11.1 classifies pumps into the more common types. For our purposes, pumps can be classified into positive displacement pumps or centrifugal pumps. Each class is used in most food plants. All food contact surfaces of food pumps should be of 304 stainless steel or equivalent.

One of the key features beyond maintenance to look for in any pump is the Cleaned In Place (CIP) feature of the pump. Tearing down pumps to clean then is costly in terms of manpower and costly

FIGURE 11.1 — Types of Pumps

Types of Positive Displacement Pumps

RECIPROCATING
Steam
  Horzontial or Vertical
    Double Acting
      Piston or Plunger
        Simplex or Duplex
Power
  Horizontial or Vertical
    Single Acting or Double Acting
      Piston or Plunger
        Simplex, Duplex or Multiplex
Controlled Volume
  Horizontal or Vertical
    Plunger or Piston or Diaphragm*
      Simplex, Duplex or Multiplex
        Manual or Automatic Control

ROTARY PUMPS
Vane
  Blade, Bucket, Roller or Slipper
Piston
  Aseal or Radial
Flexible Member
  Tube; Vane; Liner
Lobe
  Single or Multiple
Gear
  External or Internal
Circumferential Piston
  Single or Multiple
Screw
  Single or Multiple

---
* May be either Mechnanically or Hydralically Coupled

FIGURE 11.1 — Types of Pumps - Continued

Types of Kinetic Pumps

CENTRIFUGAL*
  Overhung Impeller
    Close Coupled, Single and Two Stage
      End Suction (Including Submersibles) or Inline
    Seperatley Coupled Single And Two Stage
      In Line; Frame Mounted; Counterline Support;
      Frame Mounted ANSI B73.1; Wet Pit Volute;
      Axial Flow Impeller (Propeller) Volute Type (Horz. or Vert.)
  Impeller Between Bearings
    Seperatly Coupled Single Stage
      Axial (Horz) Split Case
      Radial (Vert) Split Case
    Seperatley Coupled Multistage
      Axial (Horz) Split Case
      Radial (Vert) Split Case
  Turbine Type
    Vertical Type, Single & Multistage
      Deep Well Turbine (Including Submersibles)
      Barrel or Can Pump
      Short Setting or Closed Coupled
      Axial Flow Impellar (Propeller) or Mixed Flow Type (Horz. or Vert.)

REGENERATIVE TURBINE
  Impeller Overhang or Between Bearings
    Single Stage or Two Stage

SPECIAL EFFECT
    Reversible Centrifugal
    Rotating Casing (Pilot)

*Includes Radial, Mixed Flow and Axial Flow Designs

in damage to the impellers, vanes, etc. The cost of assembly and disassembly and the wear and tear is of major consideration when purchasing or replacing pumps. Pumps should be cleaned-in-place by pumping the cleaning solution through them, as is true of most food lines.

A pump is a mechanical device that converts mechanical forms of energy into hydraulic energy. The centrifugal pumps move liquids by increasing their speed rather than displacing or pushing them. The vanes do the work on the fluid to increase the velocity without decreasing the pressure. The centrifugal force in a centrifugal pump is the force which tends to propel outward from the center of the rotation. This increased velocity is then recovered in the casing as increased pressure.

The positive displacement pumps operate by reducing the volume of space within the pump that the liquid can occupy. In a reciprocating pump, the piston forces the liquid from the cylinder into the discharge line.

The uses for pumps go from unloading vehicles, tanks, railcars and drums, to pumping to and from various unit operations within the process, to waste water removal and aeration of waste ponds, to cleanup and hygiene maintenance by employees. Pumps are almost mandatory in the modern food plant.

In understanding pumps and pumping, there are some terms that are used in the industry that are very important. These terms are in relation to hydraulics—which is the study of fluid at rest or in motion. Understanding these terms will help the food plant operator better communicate with the pump supplier.

Absolute pressure: The sum of atmospheric pressure and gauge pressure. The absolute pressure in a perfect vacuum is zero. Absolute pressure of the atmosphere at sea level is 14.7 psi (0 psi gauge).

Atmospheric pressure: The pressure exerted by the weight of the atmosphere. The standard atmospheric pressure at sea level is 14.7 psi.

Brake horsepower: Total power required by a pump to do a specified amount of work. Brake horsepower equals hydraulic horsepower plus mechanical and other losses.

Capacity: Actual pump delivery at a specific head (in gallons per minute (gpm) or cubic centimeters per hour).

Cavitation: A result of inadequate Net Positive Suction Head Available (NPSHA). When pressure in the suction line falls below vapor pressure of the liquid, vapor is formed and moves the product with the liquid flow. These vapor bubbles, or "cavities," collapse when they reach regions of higher pressure on their way through the pump. The violent collapse of vapor bubbles forces liquid at high velocity against the metal, producing surge pressures of high intensity on small areas. These pressures can exceed the compressive strength of the metal, and actually last out particles, giving the metal a pitted appearance. The other effects of cavitation are drops in head, flow and efficiency.

Density: Density is the mass per unit volume of a substance. It is affected by the variations in gravity or acceleration.

Electrical terminology: **Ampere** is a unit of electrical current. The unit is used to specify the movement of electrical charge per unit time through a conductor. **Kilowatt** is the unit commonly used to describe electrical power. One kilowatt is equal to approximately 1.34 horsepower. **Ohm** is the unit of electrical resistance, that is, the resistance of a circuit in which a potential difference of one volt produces a current of one ampere. **Power** is the rate of doing work. **Power factor** is the percentage of apparent electrical power (volts X amps) that is actually available as usable power. **Volt** is a unit of electrical potential. A volt is the driving force which causes a current of 1 ampere to flow through a resistance of 1 ohm. **Work** is a force through a distance (ft.-lbs.)

Fluid: A substance which, when in static equilibrium, cannot sustain tangential or shear forces. This differentiates fluids from solids. However, in motion, fluids can sustain shear forces because of the property of viscosity. A fluid can be a liquid or a gas.

Head: The vertical height of a static column of liquid corresponding to the pressure of a liquid at that point. Head can also be considered as specific work (ft.-lb./lb.) necessary to increase the pressure, velocity or height of a liquid value.

Horsepower: Power delivered while doing work at the rate of 550 ft.-lbs. per second or 33,000 ft.-lbs. per minute, 0.706 BTUs per second or .746 kilowatts.

Net Positive Suction Head Available (NPSHA): The energy, above the vapor pressure of the fluid, available at the pump suction to push the fluid into the pump.

Net Positive Suction Head Required (NPSHR): The losses from the suction connection to the point in the pump at which energy is added, generally, through the impeller vanes.

Pressure: The force exerted per unit area of a fluid.

Specific gravity: The ratio of its density to that of a some standard substance. For liquid, the standard is water (1.0 specific gravity) at sea level and 60°F (see Table 11.1).

TABLE 11.1 — How Sanitary Rotary and Centrifugal Pumps Compare

Rotary Pumps	Centrifugal Pumps
1. Slow speed, with less aeration or agitation.	High speed.
2. Higher pressures–maximum 200 psi.	Comparatively low pressures– usually a maximum of 130 psi for food applications.
3. Capacity up to 200 gpm.	Capacity up to 500 gpm.
4. Some designs cannot run dry; depends on type or rotor and product.	Can run dry.
5. Temperature limitation of 220°F.	Temperature limitation of 450°F.
6. Water seals.	Water seals.
7. Uniform discharge at a given rpm.	Discharge rate varies with changes in static conditions.
8. Periodic maintenance.	Minimum maintenance.
9. Self-priming.	Requires flooded suction.
10. Relationship of pump capacity to rpm discharge is relatively linear.	Variations of pump rpm seriously affect efficiency.
11. Not shear-sensitive.	Shear-sensitive.
12. Prone to damage by foreign bodies.	
13. Vented cover available.	

PUMPS AND PUMPING

Static Discharge Head: Vertical distance from pump center line to the free surface of the liquid in a discharge tank or point of free discharge (see Figure 11.2).

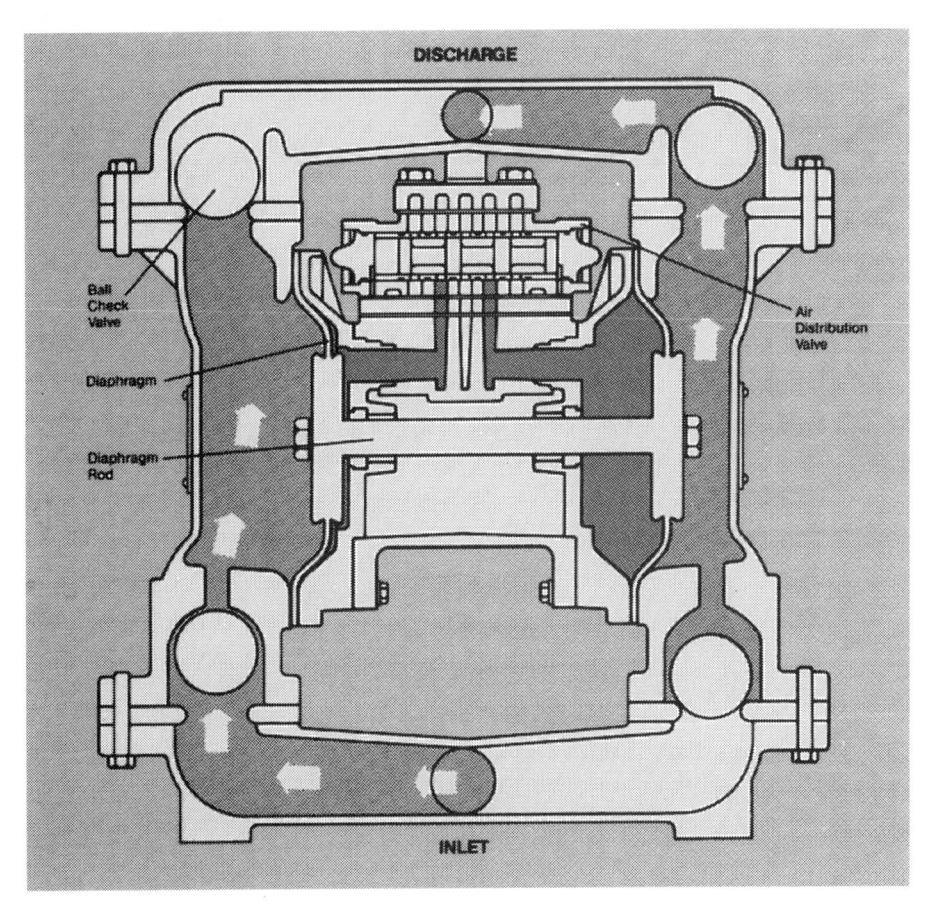

FIGURE 11.2 — Air-powered, Double-diaphram pump
(Courtesy Warren Rupp)

Static Discharge Head: The height to which liquid can be raised by a given pressure.

Static Suction Head: The vertical distance from the pump centerline to the surface of the liquid when the liquid supply is above the pump (see Figure 11.3).

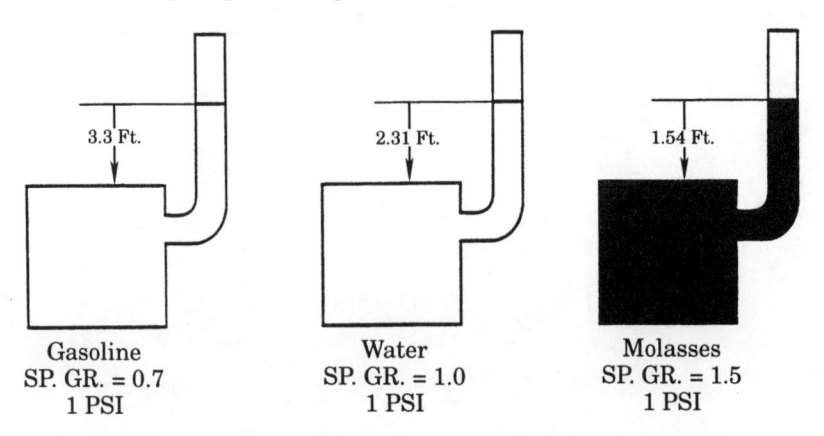

```
   3.3 Ft.              2.31 Ft.             1.54 Ft.

  Gasoline              Water               Molasses
SP. GR. = 0.7        SP. GR. = 1.0       SP. GR. = 1.5
   1 PSI                1 PSI                1 PSI
```

FIGURE 11.3 — Static Suction Head Levels for Three Products

Vapor Pressure: The pressure at which a liquid would boil at a specific temperature.

Viscosity: The existence of internal friction or the internal resistance to relative motion of the fluid particles. The viscosity of most fluids varies appreciably with changes in temperatures, whereas the influence of pressure change is usually negligible. Some liquids have a viscosity which changes with agitation. Liquids may be classified as **Dilatant**, that is, a liquid is dilatant if the viscosity increases with agitation at constant temperature (for example, candy compounds). A **Newtonian** liquid or a true fluid, if its viscosity is unaffected by agitation as long as the temperature is constant (examples are water and mineral oil). A **Thixotropic** liquid is a liquid if its viscosity decreases with agitation at a constant temperature (examples are glue and molasses).

Pumps are like any other part of the equipment in a given unit operation, and management must utilize the best system of preventive and predictive maintenance to keep the plant in full operation. Preventive maintenance focuses on timely replacement of all those parts and things about motors and pumps necessary to keep the equipment in operation all the time the factory desires to operate. Predictive maintenance, on the other hand, focuses on predicting when the pump or motor will fail and takes the corrective action needed. Corrective action is based on knowledge of the pump or motor lifecycle. With the technology available to maintenance departments today, one should be able to maintain the equipment in workable conditions at all times.

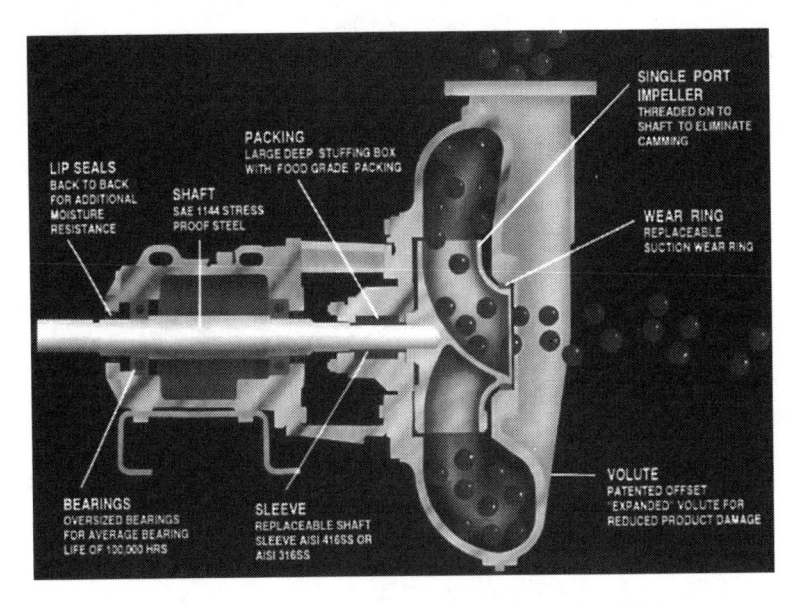

FIGURE 11.4 — Hydro Transport Food Pump
(Courtesy Cornell Pump Co.)

As stated earlier, pumps and pumping are major considerations for any food processing operation. There are many factors to consider and many choices. Every food firm should quickly learn to work with their supplier(s). Keeping this unit operation, pumps and pumping, in full readiness for operation at all times will save much in productivity and efficiency of the food plant.

Chapter 12

MIXING AND BLENDING

INTRODUCTION

Mixing and blending may be involved in the manufacture of nearly every food product. Mixing and blending are part of handling of the raw materials, sorting, and adding or combining of other ingredients and/or food additives, heating prior to filling and the actual processing to retain product quality.

DEFINITION

The concept of blending is to mingle or combine two or more ingredients. It may involve precision mixing of minute and, in some cases, trace ingredients.

OBJECTIVES

Mixing or blending of food or food ingredients is to make them more uniform or one and the same.

METHODS

One of the simplest form of mixing is running a product down a conveyor belt and dribbling on top of the product at predetermined levels from vibrator conveyors or screw conveyors or even another belt all running at an angle or over the top of the main belt one or more product(s) or flavoring ingredient(s). This has been used successfully with fruit cocktail, stewed tomatoes, peas and carrots, etc. In my opinion, it is a very successful method of mixing, in comparison to fillers for each ingredient and filling directly into the package. Generally, the latter results in layered products and may not always be highly desirable. Regardless of the system, mixing is and has been a part of this industry, and we will see many more mixes and blends in the future. Certainly, the salad business is showing the fresh fruit and vegetable industry a drastic change in merchandising by mixing the salad ingredients for the homemaker and HRI trade.

90 UNIT OPERATIONS FOR THE FOOD INDUSTRIES

FIGURE 12.1 — Jacketed Mixing Kettle (Courtesy Groen)

Kettles of various sizes, ranging from 10 to over 1000 gallon, are on the market. They may be with or without steam jackets for heating or holding while hot and with various types of stirrers, mixers with single, double or triple action agitators, or paddles (see Figures 12.1 and 12.2). The diagrammatic layouts in Figures 12.3, 12.4 & 12.5 show a tank, a tank with steam jacket, and a tank with steam jacket and mixer all in one. The descriptions are from APV Process Tank Manual.

In addition, vacuum cookers and concentrators are extensively used in the jam, jelly, sauces, catsup, and concentration of various products for bulk production.

MIXING AND BLENDING 91

FIGURE 12.2 — Jacketed Mixing Kettle Showing Agitators
(Courtesy Groen)

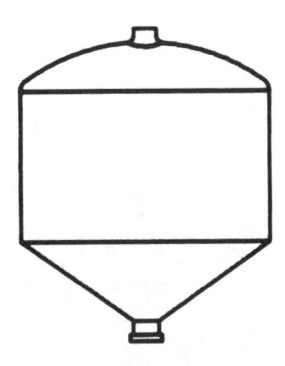

FIGURE 12.3 — Process Tank
The product will often dictate the type of basic tank configuration.
Product viscosity and the type of agitation required will determine the
shape of the tank bottom: whether it is flat-sloped, dished or coned.
The method of cleaning and manner of adding ingredients are factors
which will determine the top construction: whether you need a
full-open, hinge-cover or dome-top design. Capacity is a function of
sidewall height and tank diameter. Virtually any combination
of configurations is available, and in a wide range of capacities.
(Courtest APV Crepaco)

92   UNIT OPERATIONS FOR THE FOOD INDUSTRIES

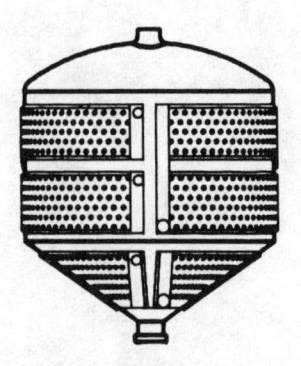

FIGURE 12.4 — Process Tank with Heat Exchange Surface
Most processing systems use some form of heat exchange.
Often, some or all of it takes place in a process tank. Dimple sheet,
headers and baffle designs provide extremely efficient heat transfer.
Add jacketed insulation and scraper blade agitation, and the time the
product is exposed to processing temperatures is greatly reduced.
(Courtesy APV Crepaco)

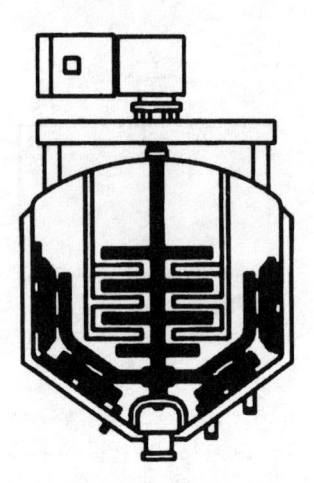

FIGURE 12.5 — Process Tank with Agitation
Specifically designed for fluid, viscous and heavy viscous products.
Today's complex formulations may dictate two, three or more types of
mixing action in a tank as the product develops.
(Courtesy APV Crepaco)

MIXING AND BLENDING

A typical batch system and a typical continuous system are shown in Figure 12.6 wherein ingredients liquid and bulk dry are mixed, blended and pumped into holding tanks for further processing.

FIGURE 12.6 — Typical Batch and Continuous Processing Systems

Typical Batch Systems

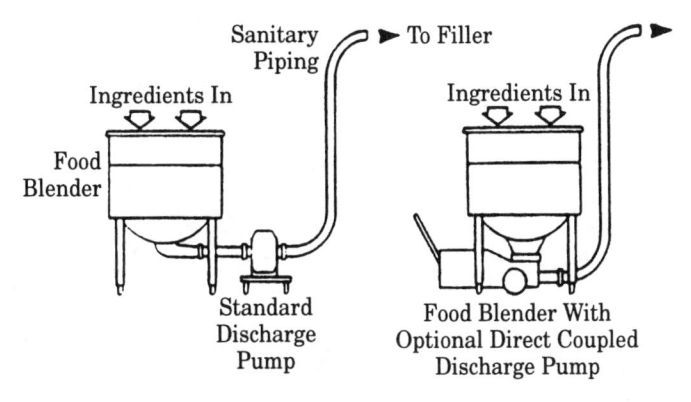

Typical Continuous Systems

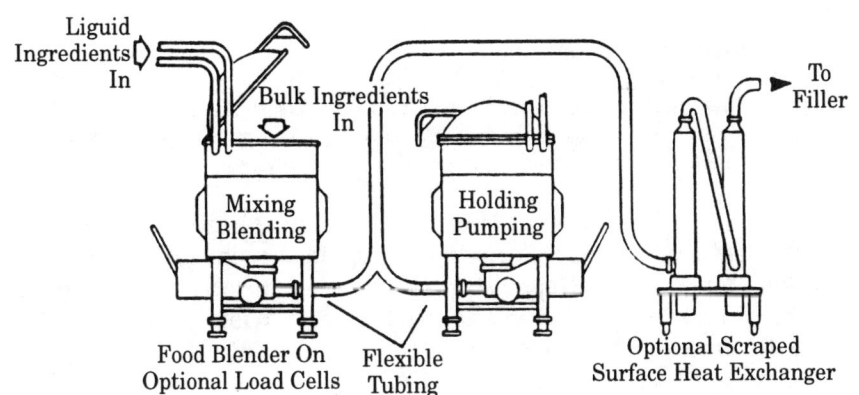

94 UNIT OPERATIONS FOR THE FOOD INDUSTRIES

A dry mixing installation and a storage (surge) tank for holding the mixed product while waiting for further processing or filling is depicted in Figure 12.7.

FIGURE 12.7 — Dry Mixing Installation and Storage Tank

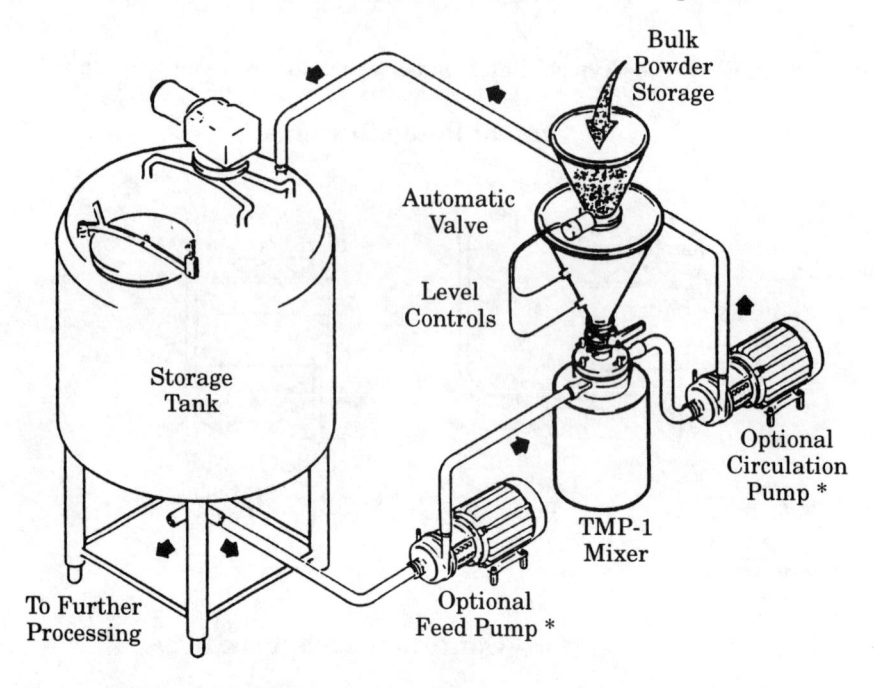

*For products with a maximum of:
- 15% total solids, pumps may not be required.
- 25% total solids, a centrifugal feed pump is required.
- 45% total solids (as shown), two centrifugal feed pumps are required.
- 60% total solids, two rotary pumps are required.

One of the key methods of mixing is using agitators. The product may be agitated when cooking the canned food as in agitated cookers. Proper agitation will speed up the cook and better preserve the nutrients and the flavor of the product. Agitation is used in many ways to create better qualities of the food we eat.

Another term used to describe agitation or blending is kneading. Kneading is the working of bread and some pastries before baking to make it work or better hold the carbon dioxide developed by the yeast prior to baking.

Mixing is concerned not only with solids, but with liquids and gases to make the product more uniform. Gases are used in packaging of foods and it is essential that they be properly mixed either in the package or before packaging. Generally, the gas already in the package is removed before the package is gassed to protect the product. Carbonation of beverages and stable foams are examples of gases being part of food items.

The mixing of liquids with other liquids and liquids with solids may be much more complicated due to the wide viscosity's of the products in question.

There are many types of mixers on the market and they all have one thing in common, that is, to mix the materials in question uniformly. Mixers can be classified as follows:
1. Flow mixers wherein two products are pumped together and the mixing is produced by interference of flow. This mixing may be further subdivided by one product being jetted into another, sparging of one product into another, circulation by producing a slow turnover of the product into another, or proportioning pumping one product directly into another.
2. Paddle or arm mixers with one or more blades, horizontal, vertical, or diagonal fastened to a horizontal, vertical or diagonal shaft and rotated on its axis within the container. The material to be mixed is pushed or moved around in swirling motions. The mixing is accomplished because the product being pushed or moved by the blades is going faster than the product between the blades. Baffles may be installed in the container to prevent stratification. Most mixer manufacturing firms have many types of mixers and they can be designed specifically for the products to be mixed once the manufacturer knows the viscosity's, consistencies, and plasticity's of the products.
3. Tumbling mixers as the name implies consist of barrels or drums wherein the product is tumbled together. Tumbling is often used for mixing powdered ingredients before adding the mixture to the product, such as, various seasonings.

In addition to the above, time and temperature may play a major role in proper mixing and blending along with the control of viscosity and consistency and plasticity of any given food. Specific equipment may be required for mixing, agitating, or stirring of the different food commodities.

Chapter 13

SALT, SALTING & BRINING / SUGAR & SYRUPING / ENROBING / BATTER & BREADING / SEASONING

INTRODUCTION
The addition of salt, syrup, batter, seasonings, vitamins, or even enrobing with chocolate is one of the so-called "chef" additions or touches to food processing. It is a technology that is ever changing and one that brings the customer back because of the flavor imparted by what is intended to be accomplished in this unit operation.

DEFINITION
One adds salt, sugar, seasonings, vitamins, etc. to foods to enhance the flavor and the acceptance of the given item. Thus, this unit operation is a means of enhancing the flavor or the overall acceptance of the product.

OBJECTIVES
To improve the flavor, the acceptance, the nutritive value, or even the appearance of a food item.

METHODS
Salt and Salting
The addition of salt or brine solutions to various vegetable products is a major aspect of flavor and texture of these food products. Small amounts of sodium chloride do accentuate the flavor and even smaller amounts of calcium chloride or other calcium compounds do materially affect the texture of some fruits and vegetables. On the other hand, too much calcium or other polyvalent ions may make a given vegetable too tough in texture. Therefore, the food processor should understand "when to salt" or "when not to salt" and in what amounts.

98 UNIT OPERATIONS FOR THE FOOD INDUSTRIES

Salt may be added directly in crystalline form, or it may be compressed into tablets and added as a tablet, or it may be made into a brine and added as part of the cover brine on canned vegetables. Many snacks have salt added in amounts varying with the commodity for example, potato chips are salted in amounts averaging 1.5% while pretzels may carry salt up to 6%. Most canned vegetables have a salt content between 0.5% and 2%. With apples, potatoes and tomatoes calcium chloride or calcium citrate, or monocalcium phosphate may be added up to 0.5%. Most baked breads and rolls, salt is added in the dough to aid in the final flavor of the product.

FIGURE 13.1 — Granular Salt Being Dispensed into Cans

Salting equipment varies from equipment that operates like a common dinner table shaker to tablet dispensers to paddle wheel and vibrator distributors and to brine dispensers as cover juices. It may be tied to a given unit operation as the salter for potato chips, or it may be part of the brine or cover juice as in canned vegetables, or it may be a stand alone unit operation as shown in Figures 13.1 and 13.2.

EXTERNAL COATINGS

FIGURE 13.2 — Salt Tablet Dispenser

Sugar and Syruping

Likewise, sugar is added to many canned and frozen fruit products to accentuate the flavor of the given commodity and to aid in prevention of discoloration. It is, also, added to remove entrapped and occluded air from the canned and frozen items. In most cases it is added as a syrup with given "put-in" values to provide given "cut-out" values (see Table 13.1)

For the canned fruits, the syrup should be added at 190°F (87°C) or above to aid in the proper formation of a vacuum in the container.

Syrups are added to the fruits in the cans just prior to closure of the container. Some use a vacuum syruper to assure the air removal before covering the fruit with the syrup. Thus, the filler may be used to add fruit to the container, vacuumize the fruit in the container, and then add the syrup to the container to cover the fruit.

TABLE 13.1 — Relationship of Put-In (P-I) Syrup vs. Cut-Out (C-O) Syrup by Label Requirements for Standardized Fruit Products
(All Data in % of Brix Values)

	Extra Heavy		Heavy		Light		Slightly Sweetened	
Product	P-I	C-O	P-I	C-O	P-I	C-O	P-I	C-O
Apricots	>55	25-40	35-55	21-25	15-30	16-21	10-15	<16
Berries								
Black	>40	24-35	30-40	19-24	20-30	14-19	10-20	<1
Blue	>40	25-35	30-40	20-25	20-30	15-20	10-20	<15
Boysen	>40	24-35	30-40	19-24	20-30	14-19	10-20	<14
Dew	>40	24-35	30-40	19-24	20-30	14-19	10-20	<14
Goose	>40	26-35	30-40	20-26	20-30	14-20	10-20	<14
Huckle	>40	25-35	30-40	20-25	20-30	15-20	10-20	<15
Logan	>40	24-35	30-40	19-24	20-30	14-19	10-20	<14
Black Rasp.	>40	27-35	30-40	20-27	20-30	14-20	10-20	<14
Red Rasp.	>40	28-35	30-40	22-28	20-30	14-22	10-20	<14
Straw	>40	27-35	30-40	19-27	20-30	14-19	10-20	<14
Young	>40	24-35	30-40	19-24	20-30	14-19	10-20	<14
Cherries, RSP	>57	28-45	34-56	22-28	25-33	18-22	10-25	<18
Cherries, Sweet	>45	25-35	30-45	20-25	15-25	16-20	10-15	<16
Figs	>40	26-35	30-40	21-26	20-30	16-21		
Fruit Cocktail	>40	22-35	36-38	18-22	30-34	14-18		
Grapes, Seedless	>40	22-35	30-40	18-22	20-30	14-18	10-20	<14
Peaches	>55	24-35	40-55	19-24	15-25	14-19	10-15	<14
Pears	>40	22-35	25-40	18-22	15-25	14-18	10-15	<14
Pineapple	>40	22-35	30-40	18-22	20-30	14-18		
Plums								
Purple	>40	26-35	30-40	21-26	20-30	18-21	10-20	<18
Others	>40	24-35	30-40	19-24	20-30	16-19	10-20	<16
Prunes	>50	30-45	40-50	24-30	30-40	20-24		

Put-In concentration must be varied depending on variety, maturity or area of production to obtain desired Cut-Out syrup concentration.

With many of the frozen fruits, the syrup is added cold or it may be added as crystalline sugar directly to the fruit either before filling or to the container with the filled fruit. Frozen fruits usually carry 2+1 to 6+1 ratios, that is, 2 parts of fruit to 1 part of sugar by weight.

Sugar is a major part of some sauces, catsups, some dressings, candies, cakes, pies, etc. In most of these items, the sugar is added as part of the formulations. Sugar is, also, used as glaze on breads and pastries and usually added after the product is baked as a flavor adjunct and as decorations. In the case of the latter it may be added by hand or by decorating machines

Enrobing

Enrobing is a major part of most candy and some cake and fruit product manufacturing. It may be accomplished by immersing a product in a tempered liquid chocolate with a finger and then removing the product with a twirl to decorate the chocolate as it is placed on a surface to cool. Apples and some fruits and vegetable products are enrobed in much the same way by using a stick in the product instead of using the individual finger. The process has been automated by coating the confectionery centers with chocolate by conveying them through a curtain of liquid chocolate followed by cooling on the same conveyor. The temperature of the food product, the temperature of the chocolate or other coating, the dwell time, and the cooling time are all critical variables that must be controlled to produce a satisfactory product.

Batter and Breading

Batter and breading of seafood for the production of fish sticks, and cakes, and portion control of cut-up fish, poultry, some types of red meat, and coating of many vegetables is becoming a big industry. It is, also, used on oysters, shrimp, scallops and many vegetables as hors d'oeuvre items. The industry is an outgrowth of what the homemaker has known for years. The commercial industry does not get the credit as it should as the technology is not as simple as dipping and rolling an item in cornmeal before pan-frying.

The batter formulation for the dip prior to the coating of bread crumbs is not a complicated subject. Batters are simply a liquid mixture of water, flour, starch, and, in some cases, seasonings into which a food is coated. The coatings may be fine or coarsely ground previously baked products. Many, variations exist in both the formulation of the batter and the breading.

Batter and breading of foods serve to enhance the food items and they give an additional flavor and texture to the given products. Batter and breaded products tend to better retain their shape during

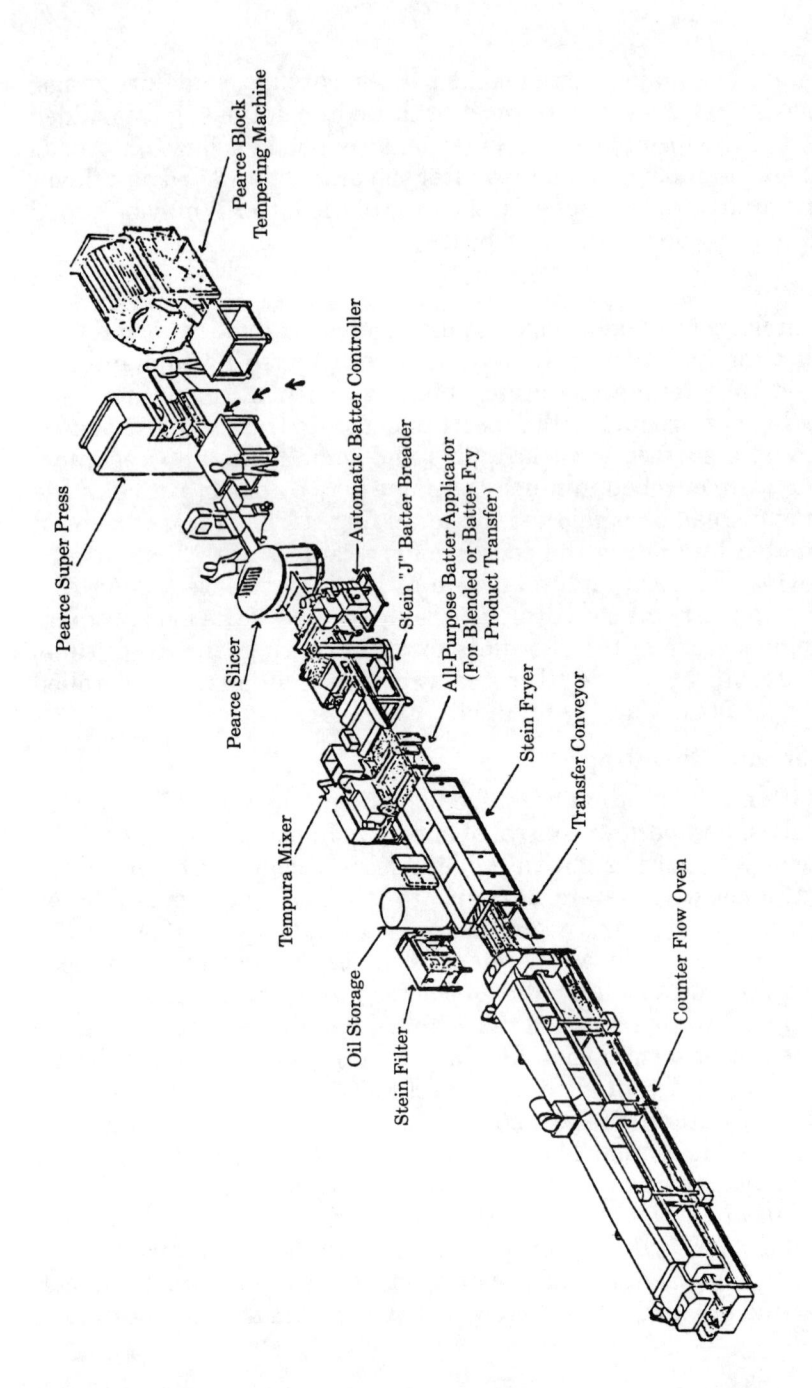

FIGURE 13.3 — Frozen Block Processing System for Poultry or Fish (Courtesy Stein)

EXTERNAL COATINGS

the processing, storage, and final cooking. The batter and breading imparts additional nutrients to the product and they prevent moisture loss. They do add weight to the product as some coatings may exceed 50%, such as, Tempura products. The coatings may present problems in the reheating and frying operations as the crumbs will increase the breakdown of the fat or oil causing off flavors if the oil is not constantly filtered to remove the "break away" crumbs. The Stein Company has spent much time in perfecting equipment for batter and breading of many food items. Figure 13.3 depicts their suggested line, along with Pearce Tempering, Pressing and Slicing for poultry products. Figure 13.4 shows Heat and Control's fryer for batter and breaded products, wherein moisture is controlled along with flavor and texture.

FIGURE 13.4 — Breaded Products Fryer (Courtesy Heat and Control)

Seasoning

Herbs, spices, onions, peppers, cheese and other approved food additives may be added to many types and classes of foods before processing, during processing and/or after processing as toppings and flavor enhancers. The equipment may involve mixing the seasoning with the main ingredient if it is to be extruded or batter and breaded or with other ingredients and added directly to the product as in sauces and stewed tomatoes. With many snack foods,

104 UNIT OPERATIONS FOR THE FOOD INDUSTRIES

the seasoning is added to the finished product directly from the cooker while the oil is still semiliquid so as to encapsulate the seasoning in the oils. Some seasonings are added to the mix at the finish of the cook cycle as with sauces and catsups.

Seasonings should be applied uniformly to the food product and they should be applied in the correct amount to produce the desired effect. The temperature of the food being seasoned may influence the point of application of the seasoning.

Figure 13.5 is an example of a topper seasoning applicator. To be effective in producing a uniform flavored product, the depth of product being seasoned will have a direct bearing on the uniformity of the seasoning being applied. Some applicators require only one layer of product under the applicator to provide uniformity of the seasoning.

FIGURE 13.5 — Roll Salter (Courtesy Heat and Control)

Further, some applicators require the product to be tumbled to assure seasoning all sides of the food item. Other applicators stress the significance of applying the seasoning as an emulsion as in an oil or cheese emulsion and spray the seasoning on the product as it is being tumbled.

FIGURE 13.6 — Tumble Drum Seasoner (Courtesy Heat and Control)

Vitamins and Other Nutrients

Some products are fortified with vitamins, minerals and/or other nutrients. They may be sprayed on the finished product directly, or added to the seasoning, or put in the mix during manufacture. These additives must be declared on the label and must be within the tolerances established by FDA. The tolerance is normally not less than 80% of the declared value nor more than 120% of the declared value.

There is no doubt that all of these food additives, from salt to nutrients to seasoning, are flavor and other quality enhancers. In many cases they are after effects of the practice of preparing and preserving the product. Some effort has been made to do a perfect job of applying these additives to the food; however, in my opinion this unit operation needs a whole lot more attention to assure the right additive in the right amount and applied at the right time. Flavor as enhanced by salt, sugar, seasoning, etc. is more than an art. Today we know the benefits of these additives and, hopefully, every unit operation operator or purveyor of this equipment will assist management in this important activity.

Chapter 14

THERMAL EXHAUSTING AND MECHANICAL VACUUM FOR CANNED FOODS

INTRODUCTION

Exhausting is a preheating operation in which filled cans or jars are exposed to heat to drive out the entrapped and occluded air from the product and the can or jar. Exhausting not only creates a vacuum which holds the lids in place on jars, but it also reduces the strain on the can or jar during the processes.

In the early history of the canning industry, it was believed that by eliminating air from the canned product and thereby producing a good vacuum, the food would be preserved. Bacteriologists have shown that there are certain types of bacteria that do not require oxygen for growth, consequently, exhausting affords no protection against anaerobic bacterial food spoilage. Nevertheless, the production of good vacuums in canned foods is needed for the recognition of good packaging.

DEFINITION

Mechanical exhausting is a means of exposing filled containers of food to heat, generally steam, to drive the air out of the container and remove the occluded air from the product to assist in creating a vacuum in the container after it is closed or sealed.

Mechanical vacuumizing of food containers is a method of withdrawing a vacuum, by mechanical means, from the container and the food prior to closing, to aid in creating a vacuum in the container.

OBJECTIVES

Bitting, Magoon and Culpepper have listed the following objectives for exhausting cans or jars prior to sealing:

1. To draw in the ends of the cans, thus giving an index to the condition of the contents,
2. To minimize the action of the contents on the container,
3. To prevent unnecessary strains on the container during processing, that is, elimination of buckling, and
4. To produce a desirable effect on the product itself.

Peterson defines the term "vacuum" as, "The pressure inside the can, as compared to atmospheric pressure, and it is usually designated in terms of inches of mercury. When air is exhausted from a container, the pressure inside becomes less than that on the outside. It is this difference in pressure that is measured by the vacuum gauge. When no differential exists between the outside and the inside of the can, the vacuum reading is zero. When complete air removal has been accomplished, the vacuum registers the barometric pressure which, in turn, is dependent on local weather conditions and latitude."

Under most conditions, air is present in canned foods prior to closure. It may be occluded within the tissues, between the tissues, and/or in the head space above the product. Products like spinach have a high amount of occluded air prior to blanching, while products packed cold in brine will usually have considerable head space air. Also, products like dry beans may have considerable air among them.

The air, in any case, should be removed to produce a vacuum within the can after the lid has been attached. The usual procedure is to exhaust the can or preheat the product prior to filling. For some products, heating is undesirable and the processor will resort to mechanical vacuum means. More recently, the injection of steam directly into the head space (steam vac or steam flow closure) has been applied. The methods to use depend somewhat upon the types of product, the operating conditions within the factory, and the desired degree of vacuum.

METHODS

There are two major types of exhaust boxes in general use, that is, the hot water exhaust box and a steam vapor box.

In the hot water type exhaust box, the cans are conveyed through the hot water. The containers are conveyed through the box on cables, discs, or chains. The water is heated externally or internally by direct steam injection. The extent of the heating is controlled by the dwell time in the hot water and, of course, the temperature of the hot water.

In the vapor type exhaust box, the heating takes place by directing the steam to the sides and bottoms of the cans or jars as they are conveyed through the box. The box may have rotary discs to move the containers through, or cables or chains to convey the cans or jars on as they travel through the chamber. As with the hot water exhaust box, the dwell time is important to assure adequate exhausting of the product.

Of course, the containers are open as they travel through either type of exhaust system. However, they should be promptly sealed so that the vapors in the head space may condense and create a vacuum during cooling of the can or jar.

A third method of obtaining a vacuum in cans or jars is to jet steam directly into the head space of the container just prior to container closure. This steam will condense, as above, and create a vacuum.

Still another practice, particularly for solid pack products, is to mechanically withdraw the air from the packed product. The lid is usually clinched first, the air is removed by a mechanical vacuum pump and the lid is sealed to the container. It is a fairly common method when packing in glass where high vacuums are needed to secure the lid properly.

It should be noted that "Steam Vac" closure has replaced most exhaust boxes and much of the mechanical vacuum systems for many of the reasons as noted in Table 14.1

TABLE 14.1 — Comparison of Methods for Producing Vacuum in Canned Foods

Factor Evaluated	Thermal Exhaust	Mechanical Vacuum	Steam Vacuum
Versatility	Good	Good	Fair
Sanitation	Fair	Good	Good
Space	Large	Moderate	None
High vacuums	Fair	Good	Good
High speeds	Fair	Good	Good
Suitability to wide range of products	Good	Good	Fair
Need for close control of head space	Not critical	Not critical	Critical
Cost to purchase	Low	Medium	Medium
Cost to maintain	High	Medium	Low

Chapter 15

FILLING AND FILLERS

INTRODUCTION

Filling is without question the most important single function in the processing of most foods in terms of need for accuracy and satisfaction to the customer. Of course, it is, also, the one aspect of most processing unit operations in terms of control over the return on the investment or "the bottom line" in terms of losses or averages. Filling and fillers, as a unit operation, are the heart and soul of most food processing operations

Every food processor should understand the significance of under or over fill in terms of dollars lost or saved (see Figure 15.1). He should understand the importance of regulatory requirements in terms of label statements. Most importantly, he should understand the impact on the customer in terms of satisfaction when purchasing their products in terms of correct fill of container.

DEFINITION

Fillers and filling are unit operations that needs to be carefully regulated and understood in terms of types of products being filled, that is, particulates versus fluids and/or the homogeneity of given products. By definition, filling is to place food into containers. Filling may be done by hand or with machine. The machines today vary widely in speed and efficiency and filling should require precision and efficiency. Today there is a great reduction in the number of sizes of containers and their is much greater standardization, at least, for cans. Plastic and glass containers and flexibles are still quite variable in terms of size of containers.

OBJECTIVES

To fill containers accurately and properly for informative labeling as to quantity actually present in the container.

112 UNIT OPERATIONS FOR THE FOOD INDUSTRIES

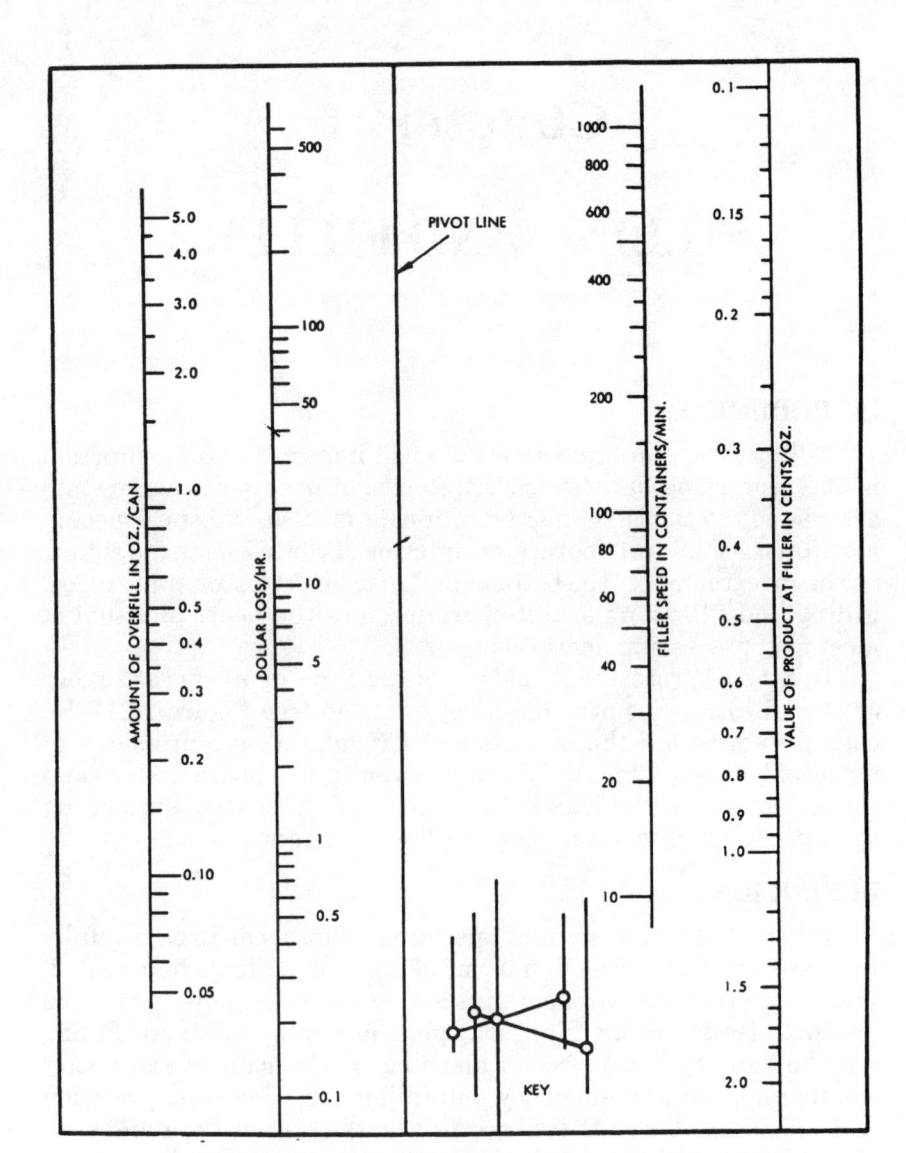

FIGURE 15.1 — With this nomograph and straight edge,
you can compute dollar loss per hour for any situation.
(Taken from M. L. Gerdes, "Four Steps to Filler Accuracy.")

METHODS

Fillers for canned and glass products vary from hand pack fillers (Figure 15.2) to gravity fillers to piston or plunger fillers to telescopic pocket and vacuum fillers (see Figures 15.3 and 15.4). They vary in speeds with the high speed fillers exceeding 1500 cans per minute on liquid items. The speed on a filler is primarily a function of the number of pockets within the machine. Normal speeds on consumer size containers vary from 100 to 300 cans per minute.

FIGURE 15.2 — Hand Pack Filler for Tomatoes (Courtesy FMC)

FIGURE 15.3 — Rotary Piston Filler (AK Robins)

114   UNIT OPERATIONS FOR THE FOOD INDUSTRIES

FIGURE 15.4 — Vacuum Filler (Courtesy AK ROBINS)

Fillers generally determine the capacity of the factory and they should always be coordinated with the closing or sealing machines. As a matter of fact, many fillers are part of the closing or container sealing system.

Aseptic type fillers for pouches and drink boxes for liquids and particulate items are becoming an important part of the industry. Speeds are now exceeding 300 containers per minute. In many cases the containers are made at the factory on demand by the filler. Blanks may be cut from printed stock, formed, filled and sealed. The container may be made from paper board, foil, or extruded polyethylene. The carton blanks are formed into containers chemo-thermically sterilized and under aseptic conditions filled with sterile products closed and ejected ready for packaging and shipment.

For example, in the Combibloc system which uses pre-fabricated cartons made from gravure printed paper board-aluminum foil and multiple extrusions of polyethylene and surlyn, the cartons are fed into the filler system, sterilized, filled and sealed. The cartons are sterilized with hot hydrogen peroxide, dried by heat from propane burners to evaporate the hydrogen peroxide, filled with the hot sterile product, injected with steam in the head space, and, the final seal is effected by the use of ultra-sonic waves.

FILLING AND FILLERS 115

With many dried products, the carton is made from rolled stock that has been printed. It is formed, filled and sealed. Three methods of bag manufacturing are shown by the James River Corporation in Figures 15.5, 15.6 and 15.7).

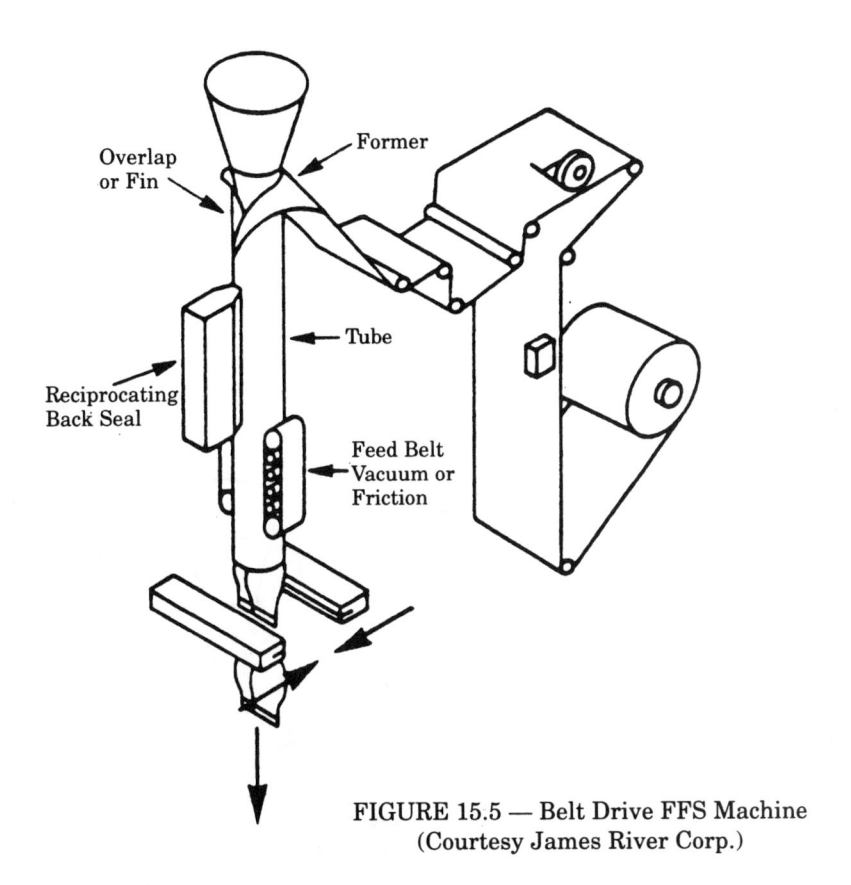

FIGURE 15.5 — Belt Drive FFS Machine
(Courtesy James River Corp.)

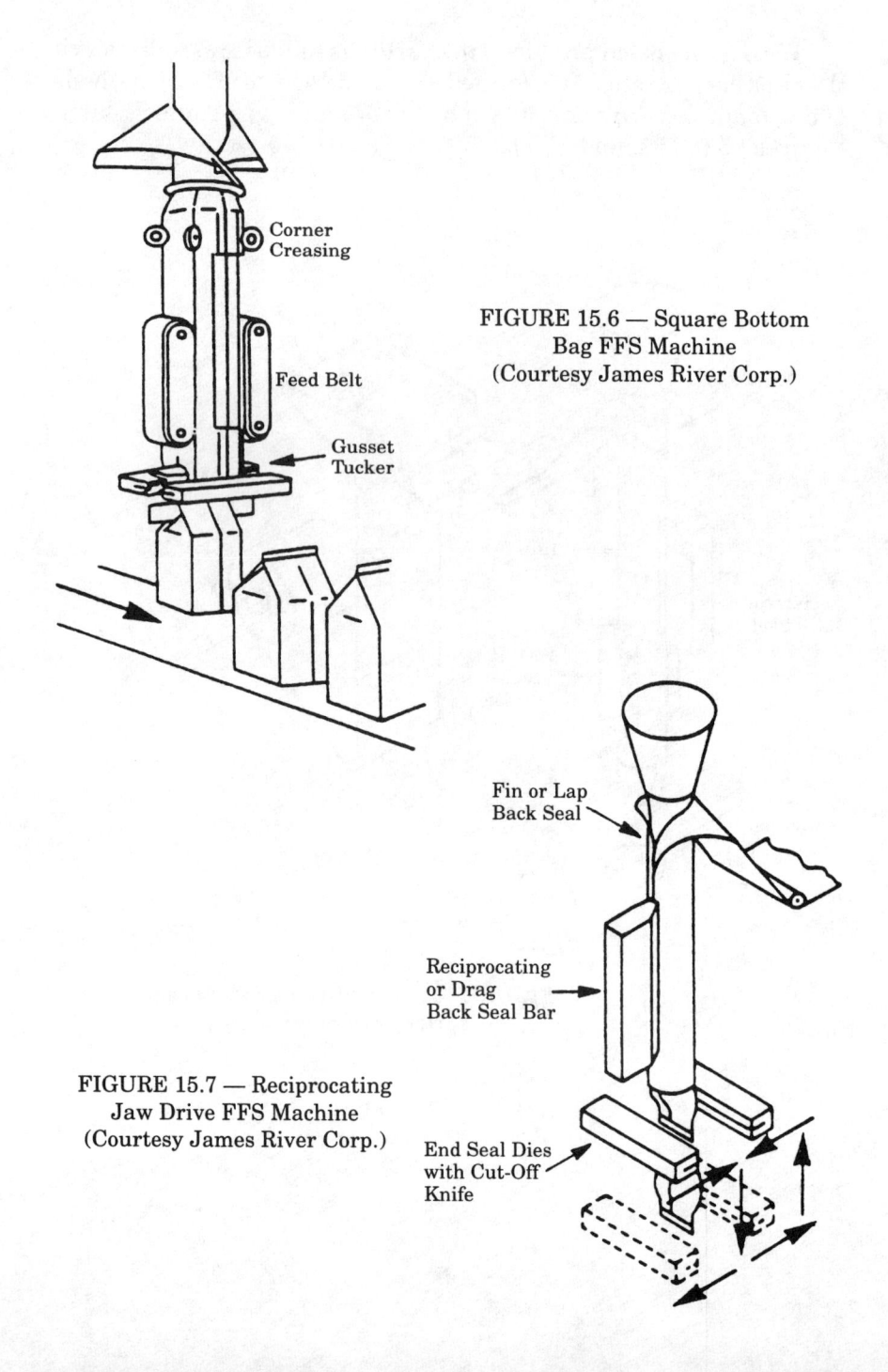

FIGURE 15.6 — Square Bottom Bag FFS Machine (Courtesy James River Corp.)

FIGURE 15.7 — Reciprocating Jaw Drive FFS Machine (Courtesy James River Corp.)

FILLING AND FILLERS

Today's fillers must be precise and accurate, high speed, sanitary in design and ease of always maintaining a sanitary condition, and must be easy to change over for different container size(s). They must be simple to operate and maintain. Most importantly, they should be economical in original cost and they should have a long life. Fillers, being the heart and soul of any food processing plant, must be respected as such.

FIGURE 15.8 — Flo-Control Overflow Briner (Courtesy AK Robins)

Chapter 16

CONTAINER CLOSURE

INTRODUCTION

Hermetically sealed containers have improved the food supply around the World for over 100 years. The security of the closure of the container is the most essential step to assure product quality and safety of the food.

Containers for processed foods are rather specific in that metal cans and glass containers are primarily used for heat processed products, plastic containers for aseptic and some hot filled items, and flexible films and aluminum foil used primarily for frozen, dried and snack items. In every packaged product, the security and integrity of the seal or the closure is the most important single aspect.

DEFINITION

Container closure is the science and technology of sealing the container securely without defects or causing further damage to the container during processing or further handling.

OBJECTIVES

The objective of container closure is to seal and/or close the container securely and precisely without any damage to the container or its seal.

METHODS

21 CFR, Part 113.60 specifically states that "regular observations shall be maintained during production runs for gross closure defects. Any such defects shall be recorded and corrective action taken and recorded. At intervals of sufficient frequency to ensure proper closure, the operator, closure supervisor, or other qualified container closure inspection person shall visually examine either the top seam of a can randomly selected from each seaming head or the closure of any other type of container being used and shall record the

120 UNIT OPERATIONS FOR THE FOOD INDUSTRIES

observations made. For double seam cans, each can should be examined for cut-over or sharpness, skidding or deadheading, false seam, droop at he crossover or lap, and condition of inside of countersink wall for evidence of broken chuck. Such measurements and recordings should be made at intervals not to exceed 30 minutes. Additional visual closure inspections shall be made immediately following a jam in a closing machine, after closing machine adjustment, or after start-up of a machine following a prolonged shutdown. All pertinent observations shall be recorded. When irregularities are found, the corrective action shall be recorded.

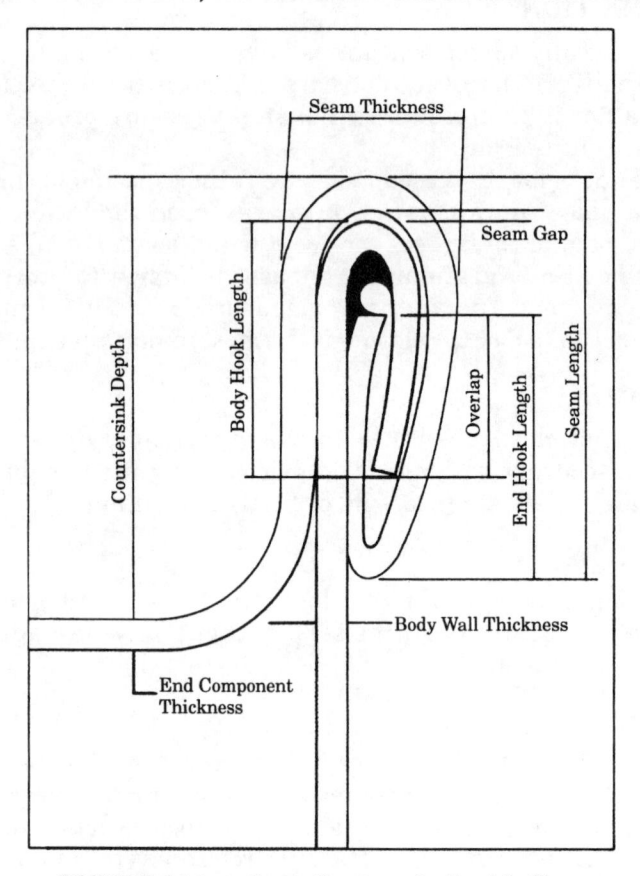

FIGURE 16.1 — Cross-Section of a Double Seam

<u>Note</u>: The Association of Official Analytical Chemists in cooperation with FDA has published some 44 pictures with details on classification of visible can defects.

"(1) Tear down examinations for double-seam cans shall be performed by a qualified individual and the results there-from shall be recorded at intervals of sufficient frequency on enough containers from each seaming station to ensure maintenance of seam integrity. Such examinations and recordings should be made at intervals not to exceed 4 hours. The results of the tear down examinations shall be recorded and the corrective action taken, if any, shall be noted.

(i) Required and optional can seam measurements:

(a) Micrometer measurement system:

Required	Optional
Cover Hook	Overlap (by calculation)
Body Hook	Countersink
Width (length, height)	
Tightness (observation for wrinkle)	
Thickness	

(b) Seam scope or projector:

Required	Optional
Body Hook	Width (length, height)
Overlap	Cover hook
Tightness	Countersink
(Observation for wrinkle)	
Thickness by micrometer	

(c) Can double-seam terminology (see Figure 16.1):

1. 'Crossover': The portion of a double seam at the lap.
2. 'Cut-over': A fracture, sharp bend, or break in the metal at the top of the inside portion of the double seam.
3. 'Deadhead': A seam which is incomplete due to chuck-spinning in the countersink.
4. 'Droop': Smooth projection of double seam below bottom of normal seam.
5. 'False seam': A small seam breakdown where the cover hook and the body hook are not overlapped.
6. 'Lap': Two thicknesses of material bonded together.

(ii) Two measurements at different locations, excluding the side seam, shall be made for each double seam characteristic if a seam scope or seam projector is used. When a micrometer is used, three measurements shall be made at points approximately 120° apart, excluding the side seam.

(iii) Overlap length can be calculated by the following formula:
The theoretical overlap length =
CH + BH + T - W, where
CH = cover hook,
BH = body hook,
T = cover thickness, and
W = seam width (height, length).

"(2) For glass containers with vacuum closures, capper efficiency must be checked by a measurement of the cold water vacuum. This shall be done before actual filling operations, and the results shall be recorded.

"(3) For closures other than double seams and glass containers, appropriate detailed inspections and tests shall be conducted by qualified personnel at intervals of sufficient frequency to ensure proper closing machine performance and consistently reliable hermetic seal production. Records of such tests shall be maintained."

Coding of containers is part of the filling and closure operation. Again quoting from CFR 113.60 Part (c) as follows:

"Each hermetically sealed container of low-acid processed food shall be marked with an identifying code that shall be permanently visible to the naked eye. When the container does not permit the code to be embossed or inked, the label may be legibly perforated or otherwise marked, if the label is securely affixed to the product container. The required identification shall identify in code the establishment where packed, the product contained therein, the year packed, the day packed, and the period during which packed. The packing period code shall be changed with sufficient frequency to enable ready identification of lots during their sale and distribution. Codes may be changed on the basis of one of the following: intervals of 4 or 5 hours; personnel shift changes; or batches, as long as the containers that constitute the batch do not extend over a period of more than one personnel shift."

Closing, capping and sealing machines are a major unit operation within this industry. The manufacturer must size the closing, capping or sealing machines to fit within the speed of the filler and other unit operations. Most packaging manufacturers lease, sell or loan these machines as part of their supplying the packages to the firm. Furthermore, the supply firms generally provide seamer, capper or sealer maintenance personnel as part of their service. However, every food firm should have one or more in-house person(s) properly trained in the maintenance and efficient running of this unit operation. These directions for closure and coding by FDA should be applicable to all processed foods, and it behooves every manufacturer to utilize the above records and codes for their firm's future. Container seam or package integrity is a must requirement in producing safe, wholesome and high-quality foods.

Coding is of particular importance to the food firm for product quality assurance and control, product shelf life, legal protection in the event of a recall, segregation of defective products, if any, answering of consumer complaints, if any, and product flow through the plant, warehouse and distribution system. The use of thermal-

FIGURE 16.2 — Ink Jet Printer (Courtesy Videojet)

124 UNIT OPERATIONS FOR THE FOOD INDUSTRIES

sensitive inks that change color from black to red, etc., and the development of video jet printing have advanced the industry greatly from the old embossing practice of yesterday (see Figure 16.2).

Bar coding is another great advance in marketing and retailing. Coding is a firm's assurance of what is happening. Codes and coding today are not only part of the original container, but also part of container casing. Firms can quickly confirm what is what, including careful inventory during manufacture.

An example code might be as follows:
 41063
 30695
Top line: 4 = Establishment, 106 = product, 3 = period of shift.
Bottom line: 306 = day of year, 95 = year.

The package should be considered an essential part of any food processing and marketing program. It is the first impression of you and your firm by the customer. It should be given a most important place in the unit operations within food manufacture and it is essential that it always be done correctly.

FIGURE 16.3 — Bar Code

Chapter 17

CANNING OR HEAT STERILIZATION AND COOLING

INTRODUCTION

Canned, glass, or pouch packed foods are sterilized by heating before filling and filling under aseptic conditions or filled into containers followed by sealing and then sterilizing by heat.

The amount of heat depends on a number of factors including: (1) particle size of the commodity (larger particles or pieces heat slower than smaller), (2) container size as all foods must be heated to the geometric center or the coldest spot in the container, (3) Consistency or viscosity of the product as thicker products heat much slower than thinner products, and (4) the pH of the given commodity. The latter is the most critical and FDA has established the dividing line between an acid food and a low acid food at a pH of 4.6.

The normal acidity of acid foods, that is, foods with a pH of less than 4.6 will inhibit the growth of public health organisms while foods with a pH of 4.6 or greater will not and, therefore, they have to be given a much more severe heat process. In the case of the low acid foods, sterilization is considered satisfactory if the pH is the equivalent of a cook of 0.7 minute or 42 seconds center temperature at 252°F or 122.4°C.

The pH of a number of foods are shown in Table 17.1. It should be clearly pointed out that mixed or blends of different commodities change the pH and the processor must know the pH to always utilize adequate and safe cooks. All process times and temperatures by container size and kind for each given commodity and process method must be on file with FDA today.

TABLE 17.1 — pH Values of Some
Commercially Canned Foods

Canned Product	Average	Minimum	Maximum
Apples	3.4	3.2	3.7
Apple Cider	3.3	3.3	3.5
Apple Sauce	3.6	3.2	4.2
Apricots	3.7	3.6	3.9
Apricots, strained	4.1	3.8	4.3
Asparagus, green	5.5	5.4	5.6
Asparagus, white	5.5	5.4	5.7
Asparagus, pureed	5.2	5.0	5.3
Beans, Baked	5.9	5.6	5.9
Beans, Green	5.4	5.2	5.7
Beans, Green, pureed	5.1	5.0	5.2
Beans, Lima	6.2	6.0	6.3
Beans, Lima, pureed	5.8	–	–
Beans, and Pork	5.6	5.0	6.0
Beans, Red Kidney	5.9	5.7	6.1
Beans, Wax	5.3	5.2	5.5
Beans, Wax, pureed	5.0	4.9	5.1
Beets	5.4	5.0	5.8
Beets, pureed	5.3	5.0	5.5
Blackberries	3.6	3.2	4.1
Blueberries	3.4	3.3	3.5
Carrots	5.2	5.0	5.4
Carrots, pureed	5.1	4.9	5.2
Cherries, black	4.0	3.9	4.1
Cherries, red sour	3.3	3.3	3.5
Cherries, Royal Ann	3.9	3.8	3.9
Cherry Juice	3.4	3.4	3.4
Corn, W.K., brine packed	6.3	6.1	6.8
Corn, cream-style	6.1	5.9	6.3
Corn, on-the-cob	6.1	6.1	6.1
Cranberry Juice	2.6	2.6	2.7
Cranberry Sauce	2.6	2.4	2.8
Figs	5.0	5.0	5.0
Gooseberries	2.9	2.8	3.2
Grapes, purple	3.1	3.1	3.1
Grape Juice	3.2	2.9	3.7

Continued on next page

TABLE 17.1 — pH Values of Some
Commercially Canned Foods, Cont'd

| | pH Values | | |
Canned Product	Average	Minimum	Maximum
Grapefruit	3.2	3.0	3.4
Grapefruit Juice	3.3	3.0	3.4
Lemon Juice	2.4	2.3	2.6
Loganberries	2.9	2.7	3.3
Mushrooms	5.8	5.8	5.9
Olives, Green	3.4	–	–
Olives, ripe	6.9	5.9	7.3
Orange Juice	3.7	3.5	4.0
Peaches	3.8	3.6	4.1
Pears, Bartlett	4.1	3.6	4.7
Peas, pureed	5.9	5.8	6.0
Peas, Alaska, (Wisc)	6.2	6.0	6.3
Peas, sweet wrinkled	6.2	5.9	6.5
Peas, pureed	5.9	5.8	6.0
Pickles, Dill	3.1	2.6	3.8
Pickles, fresh cucumber	4.4	4.4	4.4
Pickles, sour	3.1	3.1	3.1
Pickles, sweet	2.7	2.5	3.0
Pineapple, crushed	3.4	3.2	3.5
Pineapple, sliced	3.5	3.5	3.6
Pineapple, tidbits	3.5	3.4	3.7
Pineapple Juice	3.5	3.4	3.5
Plums, Green Gage	3.8	3.6	4.0
Plums, Victoria	3.0	2.8	3.1
Potatoes, Sweet	5.2	5.1	5.4
Potatoes, White	5.5	5.4	5.6
Prunes, fr. prune plums	3.7	2.5	4.2
Pumpkin	5.1	4.8	5.2
Raspberries, black	3.7	3.2	4.1
Raspberries, red	3.1	2.8	3.5
Sauerkraut	3.5	3.4	3.7
Spaghetti in Tomato Sauce	5.1	4.7	5.5
Spinach	5.4	5.1	5.9
Spinach, pureed	5.4	5.2	5.5
Strawberries	3.4	3.0	3.9
Tomatoes	4.4	4.0	4.6
Tomatoes, pureed	4.2	4.0	4.3
Tomato Juice	4.2	4.0	4.3

DEFINITION

Canning or thermal processing of food is a method of food preservation wherein a given commodity is sealed in a container (can, jar, or pouch) and sufficiently sterilized by heat to preserve it. Sometimes, canned foods are aseptically sterilized followed by aseptically filling and closing or the food may be acidified before filling or at the time of filling and then given a mild heat treatment. Canned foods are called commercially sterilized, simply meaning that all microorganisms having a public health significance have been destroyed.

OBJECTIVES

The objective of canning or thermal processing is to preserve the food and make it shelf stable and ensure its safety. *Clostridium botulinum* is the primary organism of public health significance and if the organism grows in low-acid foods, a toxin is formed which causes serious illness and death.

METHODS

All canned food preservation processes are based on specific times and temperatures, primarily developed over the years by research workers at the National Food Processors Associations, the can and glass manufacturers suppliers, food processors within the industry, and many personnel from academia. Today, these processes must be on file with the FDA. Yet the control of all processing is in the hands of the operator who must be certified by one of the many Better Process Control Schools. These schools started in 1972 at The Ohio State University and are now offered by some 20 universities with over 20,000 certified operators.

No processor should purposely exceed the cook time and temperature as over cooking will have a tendency to make the product soft with poor texture, over cooked flavor, and perhaps poor color. Processing today is an exact science and requires careful control and continuous records.

Foods may be processed aseptically, that is, sufficient heat treatment prior to filling and then filled aseptically. This practice is most common for many juices, semi-liquid products and some particulates today. The equipment is quite sophisticated and the procedures are exacting.

CANNING OR HEAT STERILIZATION

The most general type of sterilization method is to fill the container, close it, and process it in a still cooker (retort) either vertical or horizontal. This system is used for most vegetables and many fruits (Figure 17.1).

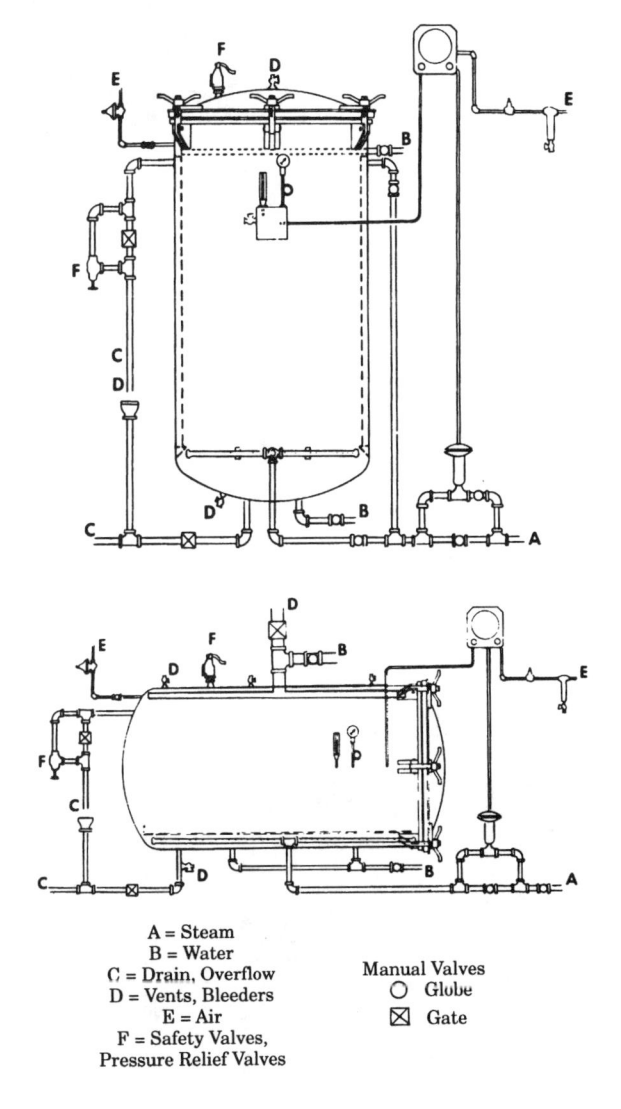

A = Steam
B = Water
C = Drain, Overflow
D = Vents, Bleeders
E = Air
F = Safety Valves, Pressure Relief Valves

Manual Valves
○ Globe
⊠ Gate

FIGURE 17.1 — Top: Vertical Retort, Bottom: Horizontal Retort. (Courtesy Food Processors Institute, Washington, D.C.)

130 UNIT OPERATIONS FOR THE FOOD INDUSTRIES

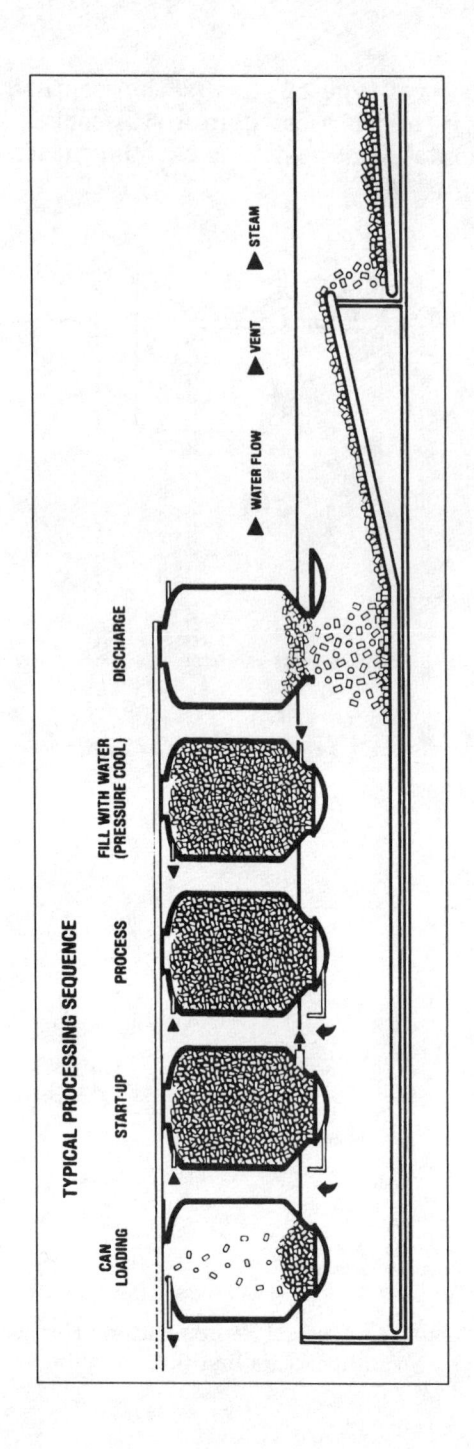

FIGURE 17.2 — Crateless Continuous Retort System (Courtesy Malo)

CANNING OR HEAT STERILIZATION

The trend, because of convenience, labor costs, and assurance of more accurate process times and temperatures, is to use continuous systems of retorting, that is, cooking as shown in Figures 17.2 using the Malo crate continuous system, or using an agitating cooker (Figure 17.3) or hydrostatic cooker (Figure 17.4, next page).

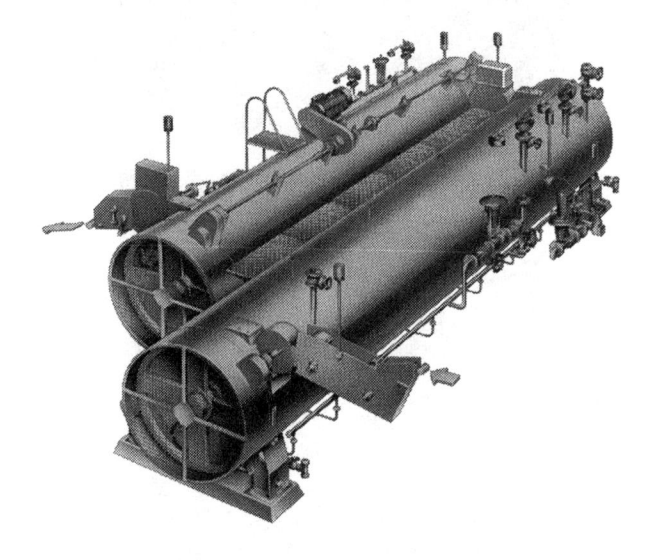

FIGURE 17.3 — Continuous Cooker/Cooler (Courtesy FMC)

If the product has low viscosity, it is homogeneous and contains no suspended solids such as juices, thin sauces, drinks and milk a plate heat exchanger system is suggested (Figure 17.5). APV states that this high ratio of heat transferred to volume ensures a low product residence time. However the system may be susceptible to fouling. Thus, the single spiral tube or tube in tube system of heat exchanger may be more beneficial (Figure 17.6). The Scraped Surface Heat Exchanger, SSHE (Figure 17.7), is useful for products having up to $3/4$ inch particulates, i.e., soups, purees and thick sauces. Heat exchangers, as here, or direct, where the product is sprayed into steam (infusion), or steam sprayed into the product (injection), have their place in the market. Of course, the steam must be potable and the product must be flash cooled so as to avoid dilution of the product with the steam.

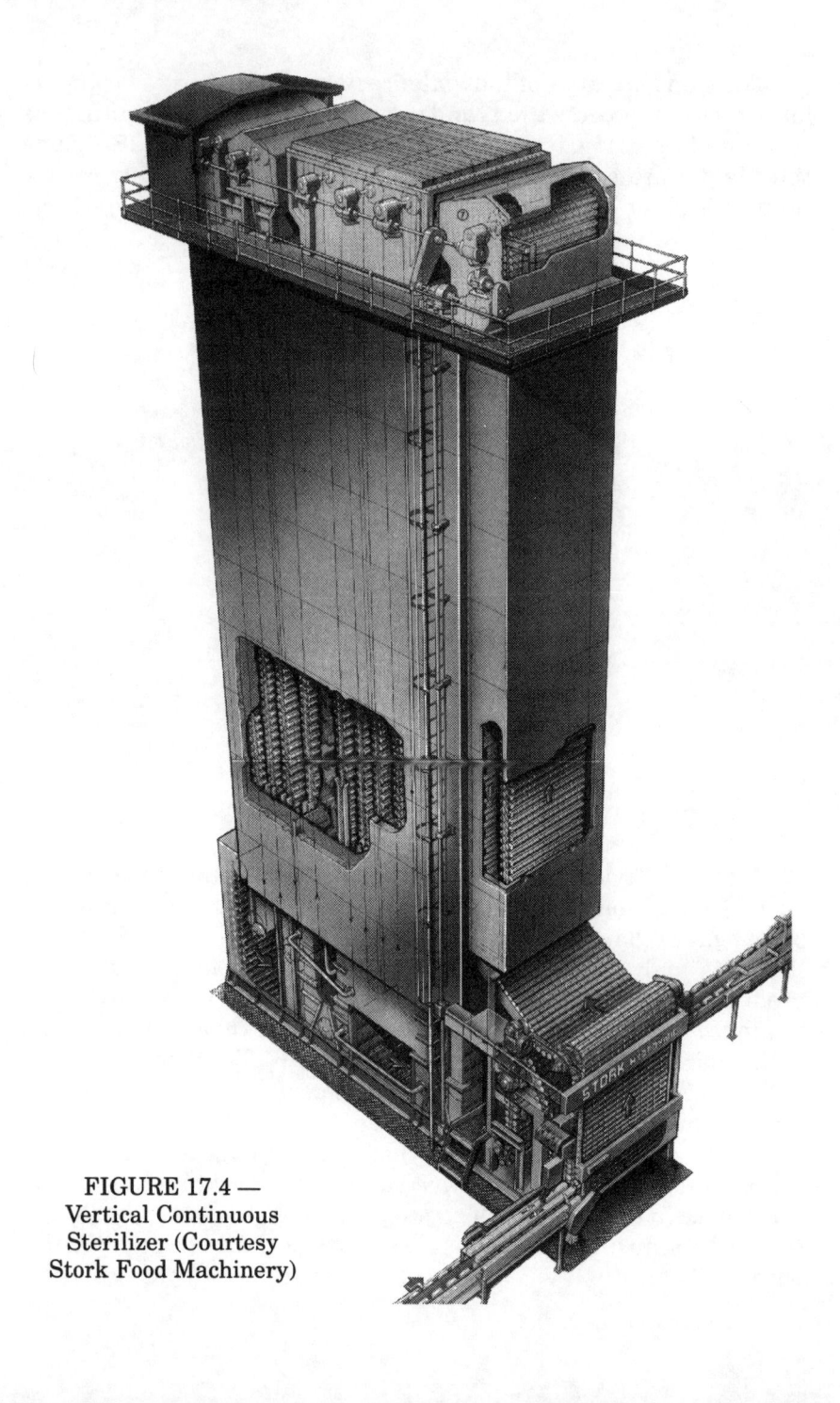

FIGURE 17.4 —
Vertical Continuous
Sterilizer (Courtesy
Stork Food Machinery)

CANNING OR HEAT STERILIZATION 133

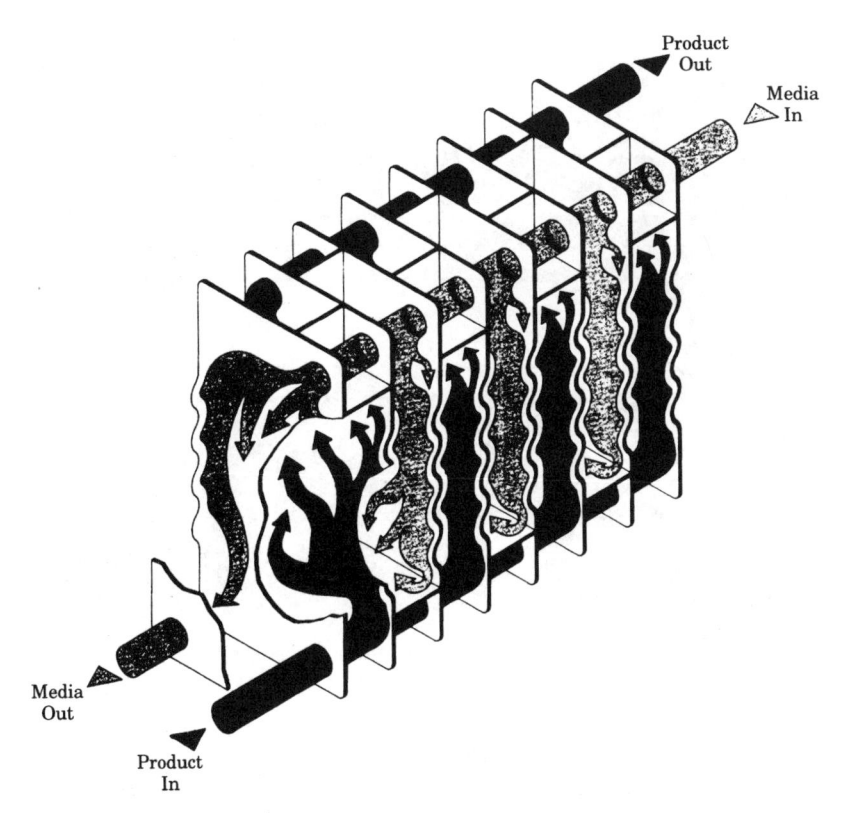

FIGURE 17.5 — Plate Heat Exchanger (Courtesy APV Crepaco)

FIGURE 17.6 — Tomato Juice Pasteurizer
(Courtesy FMC Corporation)

134 UNIT OPERATIONS FOR THE FOOD INDUSTRIES

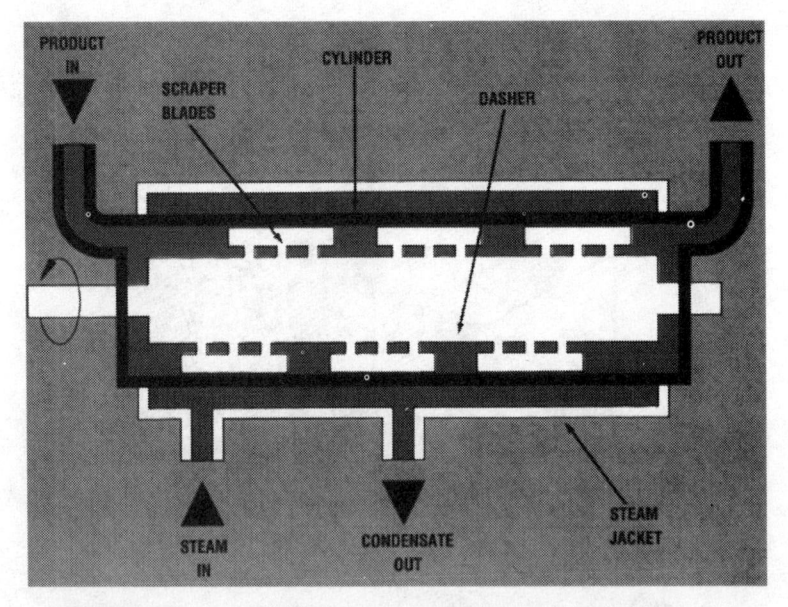

FIGURE 17.7 — Cut-away of a Scraped Surface Heat Exchanger
(Courtesy APV Crepaco, Inc.)

Some products may be concentrated under atmospheric conditions using rotary coils (Figure 17.8), or multi-stage evaporated (Figure 17.9) with even the essences recovered and added back to the concentrated product. Large concentrators are most costly, but the finished product is well worth the expense, at least, in terms of finished product quality which can be reconstituted and used in the secondary processing industry.

FIGURE 17.8 — Rotary Coils for Concentration of Tomato Pulp
(Courtesy Langsenkamp Company)

CANNING OR HEAT STERILIZATION 135

FIGURE 17.9 — Four-Stage Tomato Juice Evaporators
(Courtesy FENCO)

Cooling heat processed foods is another unit operation in the successful preservation of canned products. The object is to cool the product to a low temperature (95°F or 35°C) to 105°F or 40.6°C), but leave adequate heat in the product to dry the external moisture from the can, thus eliminating can corrosion or rusting. The length of the cooling cycle is dependent on the can size, consistency/viscosity/particle size of the product, and the temperature of the cooling water. The general practice for cooling involves immersing the cans in either: (a) a cooling canal, as shown in Figure 17.3; (b) leaving the cans in the retort after cooking, followed by adding cold water to the retort and cooling; or (c) using separate coolers, such as a spin cooler or a spray cooler (Figure 17.10). The water used to cool the cans must be free from spoilage microorganisms and there should be no chemicals in the water to cause staining on the containers. According to CFR 113.60 (b), "Container cooling water shall be chlorinated or otherwise sanitized as necessary for cooling canals and for recirculated water supplies. There should be a measurable residual of the sanitizer employed at the water discharge point of the container cooler." I interpret the latter to be greater than 3–5 ppm of sanitizer.

FIGURE 17.10 — Pasteurizer Cooler (Courtesy AK Robins)

Thermal processing, to ensure sterility and safety of processed products, is the most common practice of food preservation today. It is a safe process and the quality of the resulting products can be very high. Some improvements in equipment and process technology have been made in recent years and there is room for greater changes in the years that lie ahead. The use of acidification and formulation of products to produce acid type products (pH < 4.6) deserves much more attention than it is being given today.

Chapter 18

FREEZING

INTRODUCTION

Freezing is a method of food preservation developed in the 1920's by Clarence Birdseye, but really not commercialized until the 1930's after General Foods purchased Birdseye's patents. However, sales never really grew until after World War II after the industry learned to control enzymes and the need for blanching based on the work of Joslyn and Marsh at University of California. In the 1950's, quality was immensely improved and the industry really expanded. However, the industry did not mature until the "time-temperature tolerance" work of the researchers at USDA. Their studies placed great emphasis on handling to retain the original quality as harvested and processed. Another significant highlight in the growth of this industry was the use of evaporators and concentrators in the late 1940's and early 1950's that really opened the door for frozen concentrated fruit juices, particularly, citrus products.

DEFINITION

To understand this unit operation one needs to understand the theory of refrigeration. As Paul Christensen wrote many years ago, "refrigeration is the flow of heat. Since heat always flows from a high to a low temperature, refrigeration is the extraction of heat from the area to be cooled. Cold never moves in; heat moves out." The basis for measuring heat is the heat unit called the British Thermal Unit (BTU). The definition of a BTU is the quantity of heat required to raise one pound of water one degree Fahrenheit. The BTU as used in refrigeration occurs in large quantities and is generally converted to tons of refrigerant. A ton of refrigerant is a rate of heat extraction equivalent to melting one ton of ice in 24 hours. Since one pound of ice absorbs 144 BTU in changing from ice to water, the melting of 2,000 lbs. of ice, or one ton, would absorb 2,000 by 144, or 288,000 BTU per 24-hour day. Thus, a ton of refrigeration can be defined as a rate of heat extraction equivalent to 288,000 BTU's per day."

To make the calculations to determine the number of BTU's to freeze a given product, one must know the <u>specific heat of the food before freezing, the specific heat of the food after freezing, and the latent heat of fusion</u> (see Table 18.1).

TABLE 18.1 — Specific and Latent Heats of Foods

Foodstuff	Specific Heat (BTU/lb.) Before Freezing C_1	Specific Heat (BTU/lb.) After Freezing C_2	Latent Heat of Fusion (BTU/lb.) h_f
Asparagus	0.95	0.44	134
Berries	0.89	0.46	125
Beans, green	0.92	0.47	128
Cabbage	0.97	0.47	130
Carrots	0.87	0.45	122
Peas, green	0.80	0.42	108
Fish	0.82	0.41	105
Oysters (shelled)	0.90	0.46	124
Bacon	0.55	0.31	30
Beef, lean	0.77	0.40	100
fat	0.60	0.35	79
dried	0.34	0.26	22
Mutton	0.81	0.39	96
Poultry	0.80	0.41	90
Pork	0.60	0.38	66
Veal	0.71	0.39	91
Eggs	0.76	0.40	98
Milk	0.90	0.46	124
Water	1.00	0.53	144

Taken from *Refrigerating Data*.

The specific heat is the amount of heat that must be added or removed in order to change its temperature 1°F. The latent heat of fusion is the amount of heat that must be added or removed to change the phase of the product with no change in temperature. To change water to steam at 212°F or 100°C = 970.4 BTU, and to change water to ice requires 144 BTU's. To calculate the heat removed in freezing, the following formulae is offered for your use:

$$Q = W\{c1(t1-tf) + hf + c2(tf-t3)\}$$

Q = Amount of heat to remove
W = Weight of the food
c 1 = Specific heat of the food before freezing
c 2 = Specific heat of the food after freezing
t 1 = Temperature of the food before cooling
t 2 = Temperature of the food at freezing point
t 3 = Temperature of the food after freezing
h f = Latent heat of fusion

Example: To freeze (0°F) a 2 lb. package of shelled oysters from 55°F would require 320.6 BTU's.

Many studies have clearly shown the need for fast "quick" freezing to obtain better quality finished products. Quick freezing is defined as passing through the "zone of crystallization" within 5 to 25 minutes. This is considered the zone when ice crystals form within the product being frozen. The more rapidly a product passes through this zone, the smaller the ice crystals and upon thawing the less the leakage as cells have not been ruptured by the larger ice crystals. Time/temperature freezing rates are most important in terms of quality, particularly the texture of the product.

OBJECTIVES

To preserve food by "quick" freezing and storage of the product at 0°F (17.8°C).

METHODS OR TYPES OF FREEZERS AND FREEZING SYSTEMS

The industry uses many types of freezers and freezing systems ranging from single contact conduction plate freezers to direct immersion freezers. Freezers can be generally classified as:
(1) Still freezing in air;
(2) Air Blast or Forced Air Freezers, generally in tunnels, but may be in rooms with cold air circulating greater than 2,000 CFM;
(3) Single or Double Contact Freezers, for packaged items;
(4) Spray or Fog Freezers, with refrigerant being sprayed over the packed or loose product; and
(5) Direct or Indirect Immersion Freezers, wherein the product is immersed in the refrigerant either packed or loose.

Refrigerants vary from low pressure ammonia units to liquid nitrogen.

140 UNIT OPERATIONS FOR THE FOOD INDUSTRIES

Birdseye's early systems accomplished freezing by placing the food on a refrigerated plate and allowing the product to freeze by conduction. This was a relatively slow method of freezing. Eventually, double contact plate freezers came on the market and freezing times were reduced by more than half. All freezing rates are dependent on the thickness of the product in the package. Modern day recommendation suggests most packaged products should be less than 2 inches thick.

Indirect contact freezing by conduction have been improved by using belts. One double contact system consists of two belts, one superimposed above the other and continuously moving the compressed product through a refrigerated tunnel using calcium chloride brine at −50°F or other refrigerant sprayed from above and below the packaged product. This system is costly to maintain.

Another system is to use steel or aluminum refrigerated plates stacked in a cabinet that can be hydraulically raised or lowered to bring the plates into direct contact with the packaged food. The plates absorb the heat from the product. Figure 18.1 shows a cut-away view of a typical manual freezer. This same system has been

FIGURE 18.1 — Batch Freezer (Courtesy BOC Gases)

automated in terms of moving the product into and out of freezer with much success. This improvement eliminates manual labor for loading and unloading.

FIGURE 18.2 — Double Contact Plate Freezer (Courtesy APV Crepaco)

1. Hydraulic Pump — 2 HP for 1-3 freezers;
2. Top Pressure Plate — evenly distributes hydraulic pressure over full plate surface;
3. Connecting Linkage — sequentially separates plates during opening cycle to set station dimension;
4. Corner Headers — provide even distribution of refrigerant;
5. Refrigerant Hoses — rubber for ammonia, bronze for halocarbon, carry refrigerant between corner headers and individual plates;
6. Trays — optional for non-cartoned packages or bulk products;
7. Freezer Contact Plates — engineered with high efficiency circulation mode for uniform heat extraction over entire surface area;
8. Insulated Doors — are equipped with heavy duty hinges and frost breaker type locks.

Cryogenic or liquid nitrogen as a refrigerant for freezing is an important part of the frozen food industry today. The product may be packaged and placed on trays (Figure 18.2) or individually quick frozen (IQF) on belts and then packaged (Figure 18.3) or handled in bulk for secondary processing. The advantage of liquid nitrogen is a very rapid freezing rate and generally good preservation of quality in terms of texture, at least.

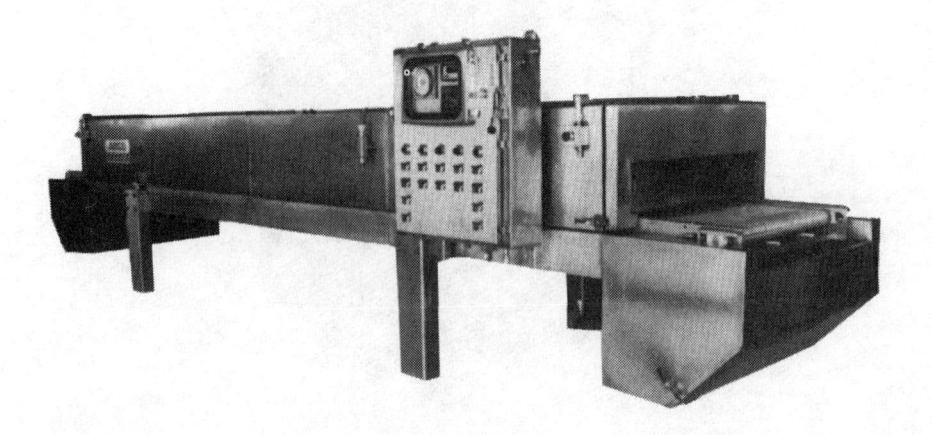

FIGURE 18.3 — Nitrogen Tunnel Freezer (Courtesy BOC Gases)

Freezing is a very successful unit operation of preserving many commodities, however, the shelf life is restricted to time. This is usually not more than one year and, of course, the product must always be kept frozen. This implies transportation and storage at temperatures below 0°F or (-17.8 C.).

Chapter 19

DRYING AND DEHYDRATION

INTRODUCTION

Drying is one of the oldest methods of food preservation known to mankind. Sun drying was used in Biblical times with up to 1 acre per 20 acres of orchard set aside for drying the fruit. During World War I commercial drying was effective in preserving many items for the military, but the product was not highly acceptable. After the work of Joslyn, Cruess, and others at the University of California on enzyme inactivation, dried products were more accepted. During World War II, dehydrated products were much in demand due to the need of food, tin was in short supply, and there was a very short rubber supply for the gasket for the tin can. The dried products had much less weight than the canned products, they had a lower bulk density than fresh or canned, they were much cheaper in price, they occupied much less space, they were considered quite good when compared to fresh that had been stored for long periods, and they generally had a fair retention of nutrients including Vitamin C. Of course, they were not as good as some would have wanted, but they did satisfy the necessary demand needed to feed a military that was on the move.

DEFINITION

Today drying or dehydration is defined as drying by artificial produced heat under carefully controlled conditions of temperature, relative humidity, and air flow. The whole purpose of drying or dehydration is to remove the moisture within the product without damage to the cells or structures of the various commodities.

There are some basic principles that should be understood when using this method of food preservation. First, **air** is used to conduct heat to the product, to carry away the liberated moisture from the product. Control of air movement means control of the rate of air drying. Second, **heat** causes the rate of evaporation of the moisture from the product. It takes approximately 1,000 BTU's of heat to change one pound of water to vapor. This is known as the Latent Heat of Evaporation. The rate of evaporation of water from a surface is directly proportional to the velocity of the air. The velocity of the air during dehydration should be from 300 CFM as a minimum to 600 CFM as optimum.

The third fundamental consideration is **relative humidity**. The relative humidity (RH) of air may be defined as its amount of saturation with moisture vapor. This is referred to as a percent of saturation, that is, air completely saturated with water vapor is 100% saturated. The absolute amount of water vapor that air can absorb, approximately doubles with each 27°F (−2.8°C) rise in temperature. If the RH is too low and the temperature so high that the moisture is removed too rapidly or more rapidly from the surface than it diffuses from the interior of the product, the outer surface may case harden. Therefore, in drying the rate of evaporation is retarded by control of the RH. Thus, controlling the rate of diffusion of water uniformly from the tissues allows a product to dry without case hardening. Another term is **critical temperature**. Every food item has its own critical temperature. This is the temperature at which when the tissues are almost dry whereby it may undergo undesirable changes in color and or flavor.

To better understand the above terms and principles of drying, the information shown in Figure 19.1 should be helpful. The important point to always remember is the critical temperature which must be determined by trial and error for each commodity. Ideally, the moisture reduction should be as rapidly as possible without getting case hardening of the product.

OBJECTIVES

To preserve food by drying or dehydration and to ensure a satisfactory product as to quality, nutrient retention, lowering of bulk density and overall acceptance.

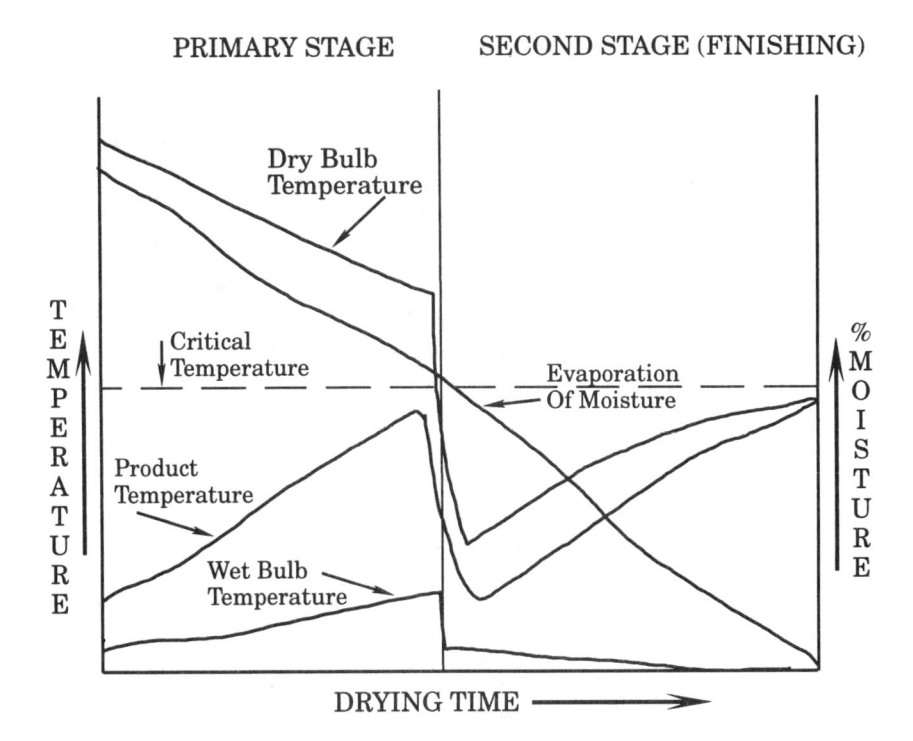

FIGURE 19.1 — Ideal Conditions within a Dehydrator under a Two-Stage Flow Dryer

METHODS

There are many forms of commercial dehydrators in use today. Generally they can be divided into (1) cabinet dryers, (2) tunnel dryers (3) spray dryers, and (4) drum dryers. Vacuum and foam mat dryers are used for some products.

Cabinet dryers are relatively simple in design. They are nearly always made for use of trays and the loaded trays stacked in the dryer. Inside the dryer there are vents to control the direction of air movement, a fan to move the air through, over, or above the product and a heat source. The heat may come from steam piped to coils within the cabinet or the heat may be generated outside and brought into the cabinet through ducts. All cabinet dryers should have a means of controlling the relative humidity either by louvers or by direct admission of steam to the cabinet. This system of drying is relatively inexpensive, but very slow in terms of production of quality dried items.

Tunnel dryers are an outgrowth of World War II, particularly in the Pacific Northwest. The principle is the same as the cabinet dryer, but much greater volume can be dried depending, of course, on the length of the tunnel. The tunnel may use a truck loaded with trays of product or, today, belts carrying the product through the dryers. Tunnel dryers may be counter flow, parallel flow or a combination, that is, the product is going in opposite direction of the air movement, or in the same direction, or parallel flow at the start and counter flow at the end. The latter is a good system. With belt tunnels, air may go across the product, through the product, or against the product. Belt tunnels have more versatility in terms of drying efficiencies and productivity's (see Figures 19.2 and 19.3).

From Van Arsdel (1951B)

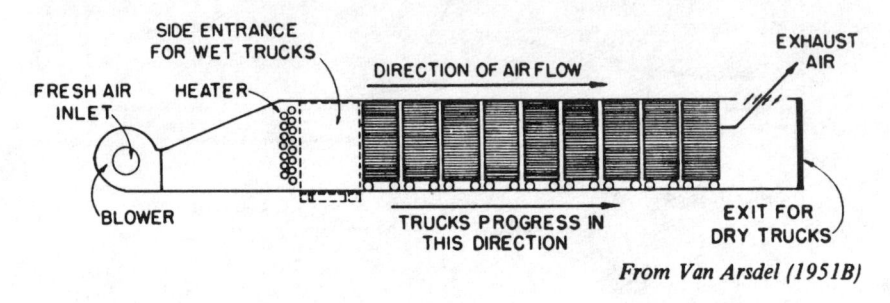

FIGURE 19.2 — Simple Concurrent Tunnel (Elevation)

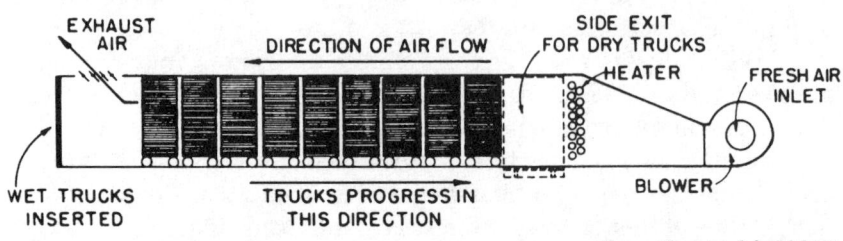

From Van Arsdel (1951B)

FIGURE 19.3 — Simple Counterflow Tunnel (Elevation)

Drum drying is an effective method for drying slurries and semi-liquid products, such as, sauces, mashed potatoes, and some soups (see Figure 19.4).

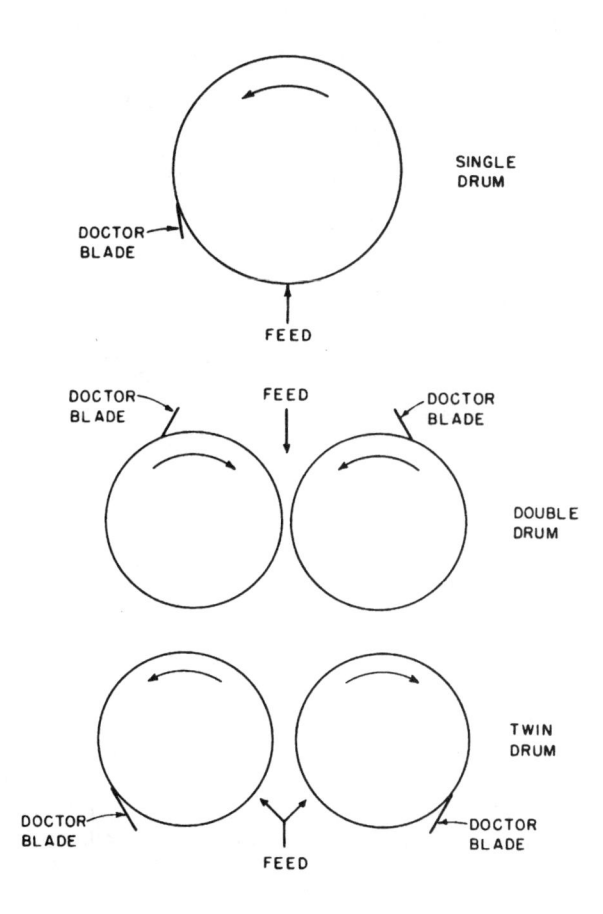

FIGURE 19.4 — Types of Drum Driers
(from Van Arsdel)

148 UNIT OPERATIONS FOR THE FOOD INDUSTRIES

Spray drying is a method of drying liquids and homogenized products. This method is particularly adapted to milk, eggs, and many fruit juices (See Figure 19.5).

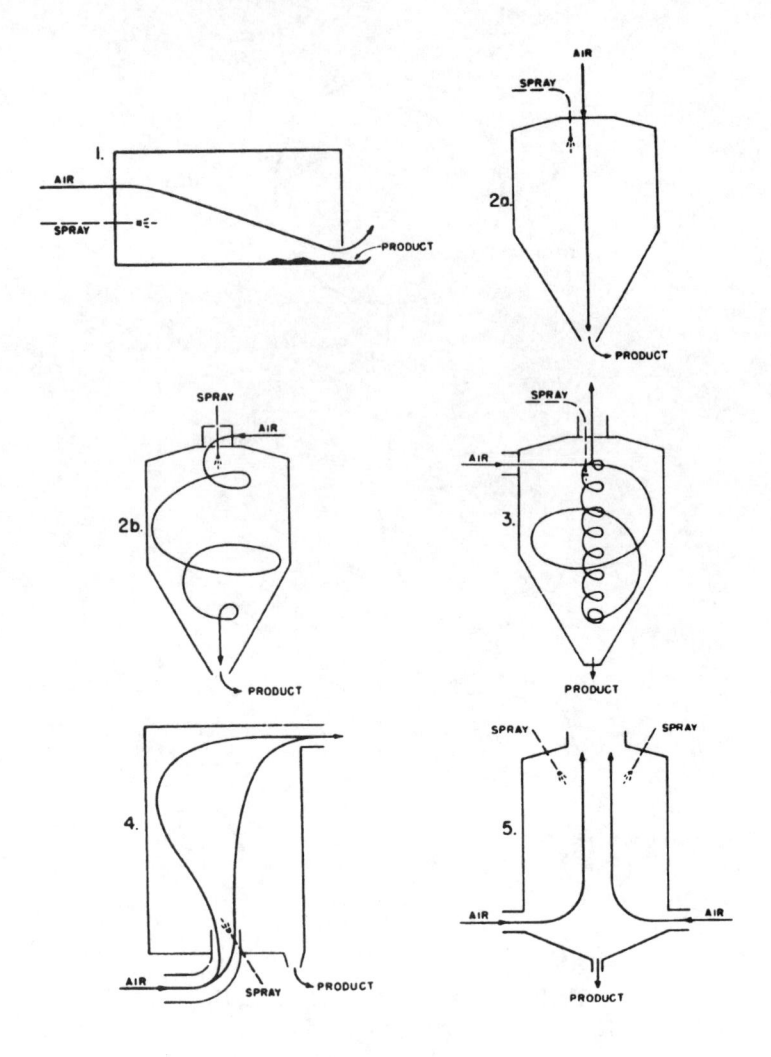

FIGURE 19.5 — Types of Spray Driers
(from Van Arsdel)

USDA personnel developed a Crater Foam-Mat dryer for paste, baby foods, and pureed products. The outstanding characteristic of the dried products is their fine foam structure so that reconstitution is extremely rapid.

Freezing products and then drying by sublimation is an important part of the dried foods industry today. These products differ from frozen products in that they are shelf stable. They differ from de-hydrofrozen products in that their moisture content is similar to dried products while dehydrofrozen products have about one-half the normal moisture and they must be kept frozen.

Chapter 20

FRYING

INTRODUCTION

Frying is an important unit operation in the manufacture of many snack foods, frozen French fries and for the preservation and cooking of many other foods. This industry has had a phenomenal growth in the past 50 years and is a significant part of the food preservation industry.

DEFINITION

Frying is a form of drying or removal of moisture from the raw food. The moisture for shelf stable items like snacks is removed to 1.5 to 2%, a reduction of some 75 to 90% of the water from the original raw product. This unit operation utilizes oil to cook in as the temperature can be elevated to nearly 400°F (204°C) and the cook time becomes a matter of seconds for some products like chips. The oil is the heart of the operation and it must be kept clean and free from degradation.

Degradation causes the oil to become rancid, that is, the onset or development of free fatty acids. Degradation causes the oil to develop off-flavors and off-odors. In addition, degradation causes the oil to become very dark in color and most unattractive. The degradation develops because of free polar materials in the oil coming from metal ions, alkaline materials and food particles. There are two ways of preventing the throwing away of oil. First, all fryers should be designed so that the oil is constantly being changed, that is, turning it over every 24 hours. Oversize fryers are, therefore, the first cause of oil breakdown. Second, the oil must always be filtered to remove any particles, polar materials, or other substances in the oil that cause it to go bad. Heat and Control® have a patented system that works to keep the oil clean (see Figure 20.1).

152 UNIT OPERATIONS FOR THE FOOD INDUSTRIES

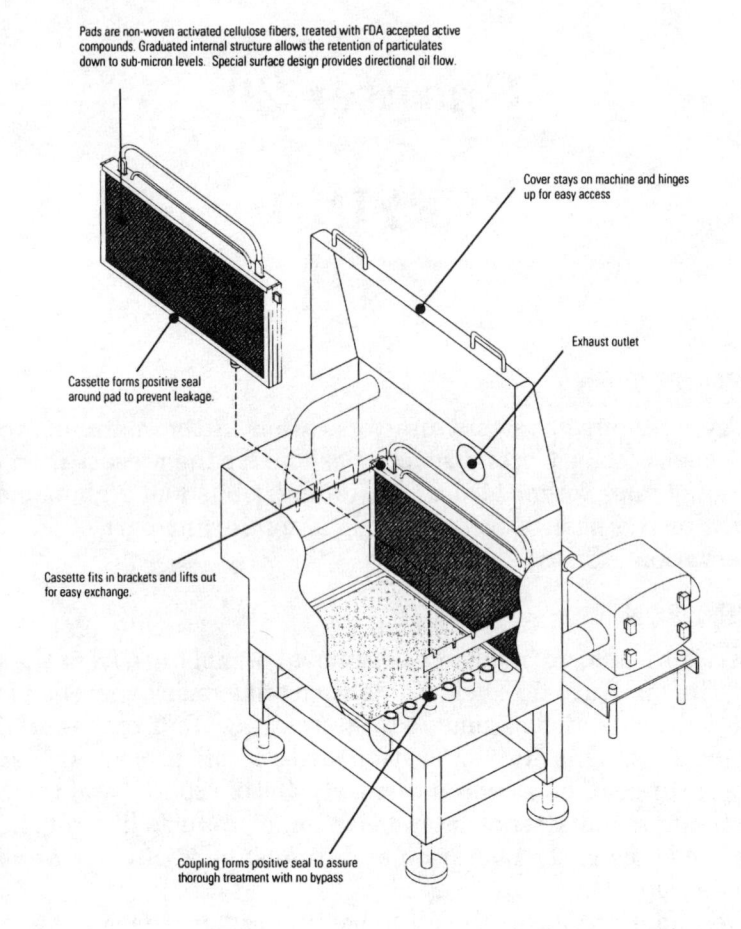

FIGURE 20.1 — Oil Treatment System (Courtesy Heat and Control)

The second key to any frying operation is the heating system to keep the oil hot for the actual processing of the commodity. Fryers are sometimes heated by direct fire with the fire in tubes or directly under the oil tank.

Most fry manufacturers now use side or separate heaters (Figure 20.2) to heat the oil and then pump it to the fryer or to separate zones of the fryer for greater efficiency.

FIGURE 20.2 — Pollution Control Heat Exchanger
(Courtesy Heat and Control)

OBJECTIVES

To make a shelf stable item that can be packaged in moisture vapor proof containers for retention of product quality. The shelf stability of the product is in great part restricted by the oil adsorbed during the cooking and frying operation.

154 UNIT OPERATIONS FOR THE FOOD INDUSTRIES

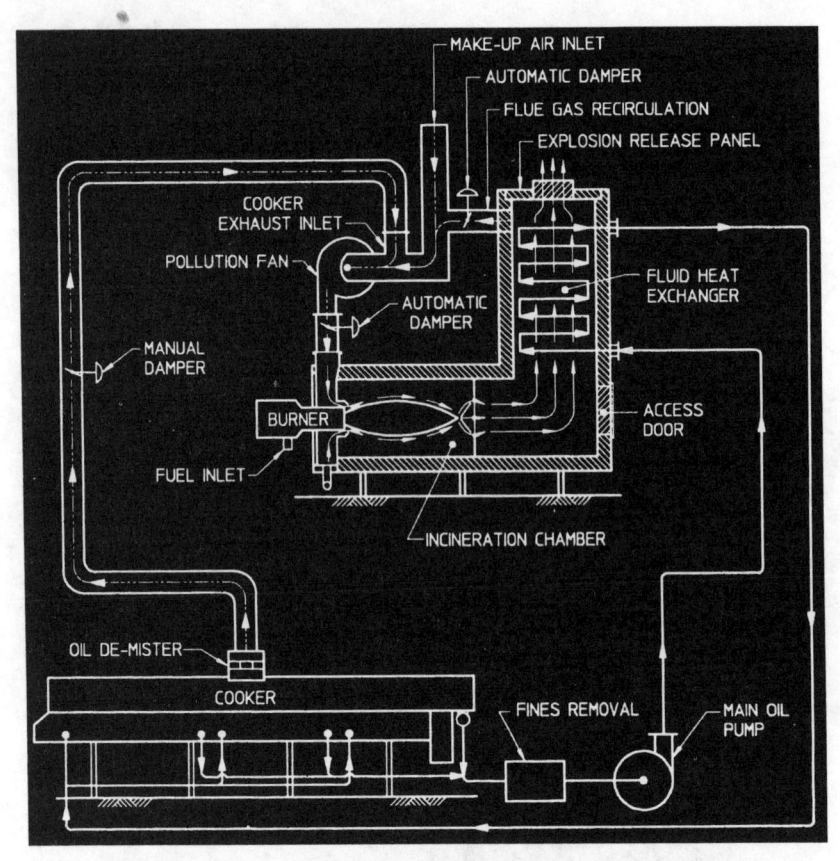

FIGURE 20.3 — Schematic of Pollution Control Exchanger
(Courtesy Heat and Control)

METHODS

Raw foods are prepared for consumption, except that they generally are sliced, diced, stripped, etc. to make them in smaller forms or shapes for quicker heating and moisture removal. After preparation, they are immersed in hot oil and fried for given periods of time to reduce the moisture for shelf stability. This moisture is usually reduced to 1.5 to 2% prior to seasoning and packaging in moisture vapor proof containers.

Fryers are of many sizes. They may be batch (Figure 20.4), semi-batch, or continuous (Figure 20.5). Continuous fryers have variable speed drives and should have multiple zones for the manufacture of high quality products. Some firms may use two continuous fryers operating in tandem to compensate for the temperature differences needed.

FRYING 155

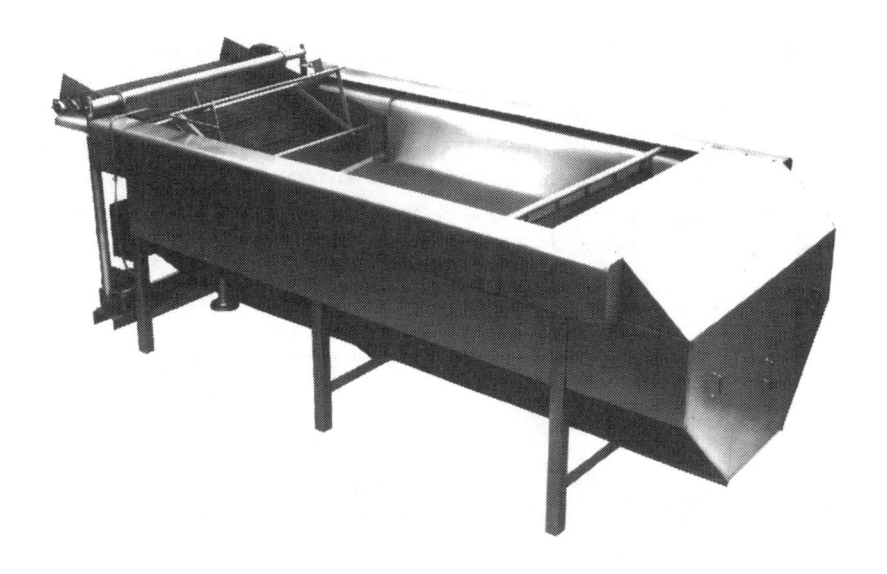

FIGURE 20.4 — Batch Fryer (Courtesy Heat and Control)

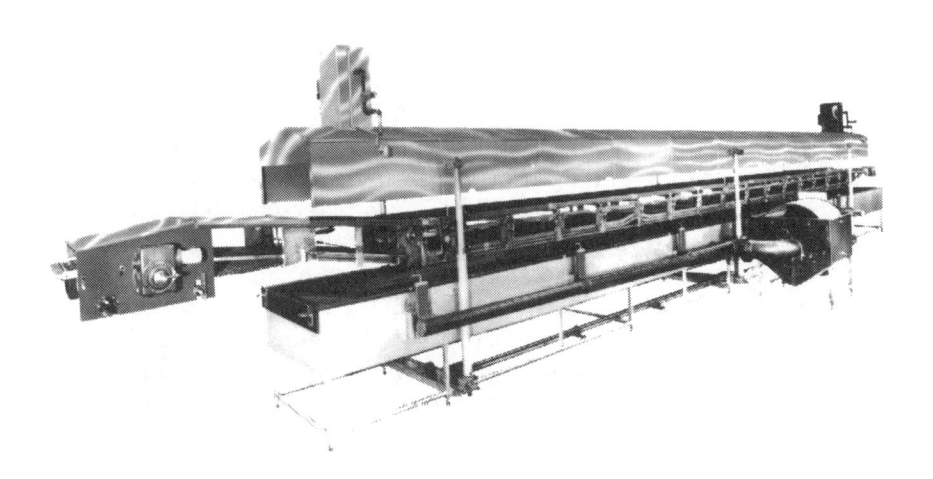

FIGURE 20.5 — Continuous Formed Potato Products Fryer (Courtesy Heat and Control)

156 UNIT OPERATIONS FOR THE FOOD INDUSTRIES

In many parts of the world, the food is shipped in as dried pellets high in protein and/or carbohydrates. Pellets can be stored for long periods of time under wide range of conditions. The pellets are then popped (Figure 20.6) in oil and become a good source of very high energy foods. They may lack certain nutrients, but if properly supplemented popped pellets are most suitable. In the U.S., pellet popping is a big business and the food products are in demand.

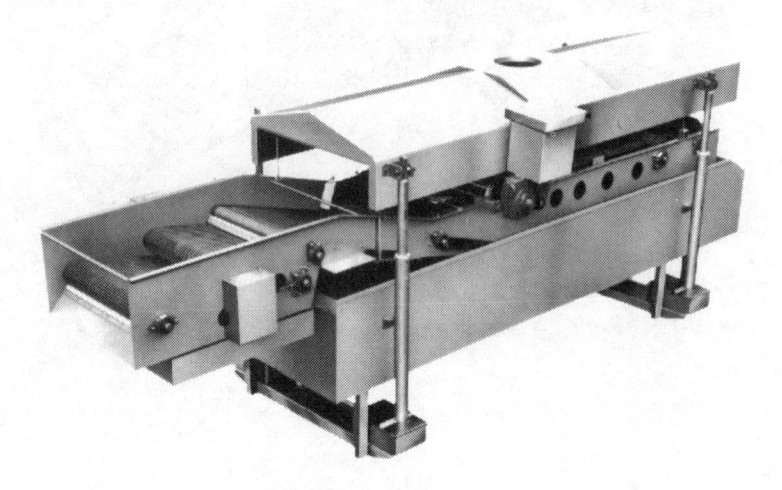

FIGURE 20.6 — Pellet Popper (Courtesy Heat and Control)

Regardless of the type of fryer, the oil in the fryer should be continuously filtered and when not in use, it should be cooled as quickly as possible and covered with nitrogen to keep the air away from the oil, thus preventing the onset of rancidity. The oil should never be thrown away if the fryer is properly sized.

The control of oil adsorption is a function of moisture content in the product being fried, the thickness of the individual piece of food, and the time and temperature the product is subjected to. Processors should remove as much water as possible from the raw product before the product enters the fryer. Secondly, the greater the surface area of the food particle, the greater the oil adsorption. Thirdly, the lower the oil temperature, the longer the dwell time and consequently the greater the oil adsorption. Oil levels in foods can be controlled and they should not exceed 30% of the weight of the food item.

FRYING

Thus, sizing the fryer to the job and operating at desired temperatures with short dwell times in clean oil will assure the growth of this method of food preservation.

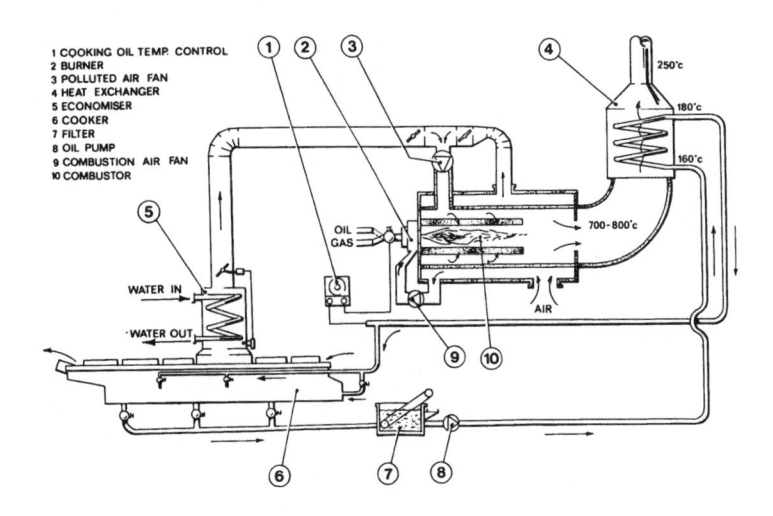

FIGURE 20.7 — Schematic of Heat Exchanger For Heating Wash Water or Specific Rooms (Courtesy Flo-Mech).

Chapter 21

EXTRUSION COOKING

INTRODUCTION

Extrusion, as a unit operation in food preservation, is an outgrowth of the late 1930's. It is a unit operation that forces a food material (wheat, corn, rice, potatoes, oats, etc.), with a moisture content between 20 and 40%, through a die under a condition of varying time, temperature, pressure, shear, and mixing. The first extruded products were pasta items. Today, many snacks and a wide array of foods are extruded.

DEFINITION

Extruded foods have been defined as a High Temperature Short Time (HTST) method of preservation. Extruded foods come in many shapes and sizes, depending on the die and operating parameters of the extruder. The shapes of extruded products range from curls, balls, stars, ovals, rounds, shells, chips, flakes, hats, and others, and the only shape limitation is due to man's ingenuity.

Extruded foods have their own texture. The texture is primarily controlled by the moisture content left in the product during the extrusion process.

The flavor of extruded products can be controlled and new flavors are easily developed using heat stable flavors or seasonings.

Extrusion of foods aids in making the extruded food more digestible, as the starch granule is completely ruptured and gelatinized by the combination of moisture, heat, pressure, and mechanical shear. Due to the high temperature used in extruding many products, the food is completely sterilized, cooked, and ready for consumption.

OBJECTIVES

To produce food products by HTST cooking to gelatinize the starch to make the product more digestible by man. Also, to manufacture shelf-stable foods with high nutrient retention.

METHODS

According to J. M. Harper, "Extruders consist of a flighted Archimedes screw which rotates in a tightly fitted stationary sleeve or barrel. The action of the flights on the screw pushes the food product forward, and, in doing so, mechanically works and mixes the raw food mass. Heat is normally added to the food through the viscous dissipation of the input energy, or by heat transfer from heating or cooling jackets which surround the barrel." Jack L. Rosen and Robert C. Miller describe the screw of an extruder. They state that the "screw is divided into several sections, each with a specific purpose. The feed section picks up the food from the hopper and propels it into the main part of the extruder. In the compression zone, the loosely packed food material is compacted and converted from a flowing granular or sticky mass to a relatively uniform plasticized dough. Some external heat must be applied but most of the heat is generated by friction. The metering section contributes to a uniform flow rate required to produce uniform dimensions in the finished product, and build up sufficient pressure in the extruded material to force the plastic dough through the rest of the extruder and out the die." This metering section is designed to thoroughly mix and/or increase the temperature of the extruded material.

There are many types of food extruders, including single screw, twin screw, forming and high screw, and others. Figure 21.1 shows a cross-section of a typical food extruder.

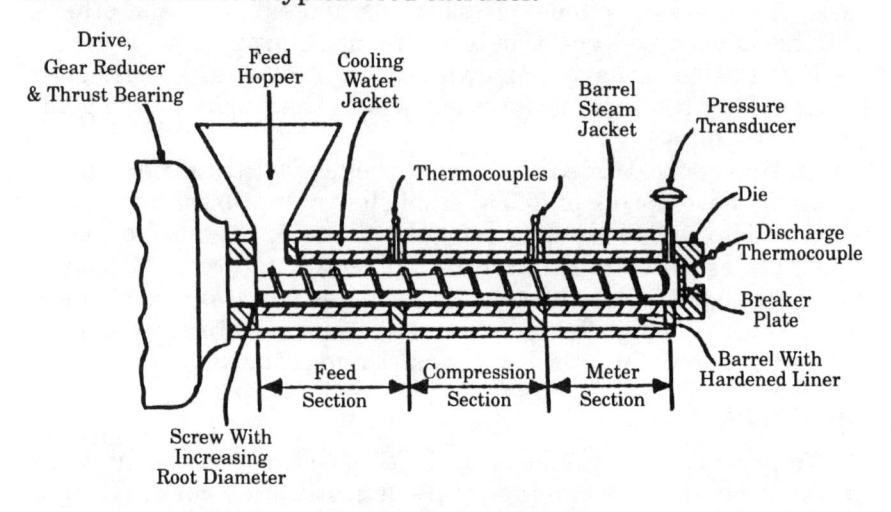

FIGURE 21.1 — Cross-section of a Typical Cooking Food Extruder

A method of classifying extruders can be based on their relative shear and pressure. Products produced with a low pressure and high shear are called collets. At the other extreme, products produced with high pressure and low shear are called pastas. In between, there are many variations of extruded products, depending on the pressure and shear used in the extruder. Of course, temperature and moisture have a whole lot to do with the texture of the extrudate. Extruded products vary from cereals to pretzels. Because of the simplicity, the high rates of production, the quality, and the great return on the dollar invested, I believe this part of the industry is still in its infancy and about to explode. It's taken mankind time to accept the texture, and it's now time to accept the flavor and the great value of these food items. The industry needs to continually control the quality in order to see this industry really "take off."

Chapter 22

ASSURING THE SAFETY OF OUR FOOD

Since the turn of the 20th Century, food processors have always striven for the production and manufacture of safe and wholesome foods. It is still a full time requirement and can only be successful with well defined programs using continuous surveillance with update teams practicing modern communications.

Responsibilities for food safety start with management, as food safety is a management function. The authority for carrying out in-plant safety functions must be assigned responsible persons.

Food safety requires good records covering all facets from procurement through preparation, processing, and packaging, including the human factor. The records must be accurate and reflect the actual operating conditions at the time specified. The records must be in ink and the process operator(s) must sign or initial as the records are developed. Further, all records must always be on appropriate forms.

The FDA and USDA, working in concert with NFPA and many other bodies, have developed a NEW system known as Hazard Analysis Critical Control Point or HACCP. This system has been defined as the identification of sensitive ingredients, critical process points and control limits, and relevant human factors as they affect product safety. The technique maximizes the beneficial impacts of management decisions by first identifying potential product risks. Those risks identified as critical are then evaluated and strategically located control points are established to monitor these product and process risks to assure the safety of the food. Thus, HACCP is preventive in nature and it protects the consumers from exposure to potential food hazards.

Penalties for failure in food safety are most severe: They include both fines and imprisonment.

HACCP is a management tool for assuring food safety. It should be the top priority of any food firm. HACCP requires that everyone in management work together to identify all potential hazards and risks in the operation and develop appropriate prevention programs to control the process and apply all corrective actions where needed.

HACCP requires training of all personnel to understand identification and monitoring requirements. Management and all employees must train together to design and control the process through identification of potential hazards and how to prevent the system from going wrong. A strong HACCP training program followed by empowerment of manufacturing employees will provide motivation for success.

HACCP requires a strong commitment by management with a thorough communication up and down the line. HACCP requires the right product design and process control plan(s) using a multidiscipline team (personnel from procurement, production, engineering, quality assurance, sanitation, and management) that truly embrace the concept of HACCP. Employees must want to take ownership of the process for HACCP to be beneficial and successful.

HACCP HAS SEVEN CLEARLY DEFINED STEPS:

1. Identification of all the HAZARDS associated with the product and the process.
2. Identification of the CRITICAL CONTROL POINTS (CCP) in the process.
3. Establishment of the CRITICAL LIMITS (CL) for preventive measures associated with each identified CCP's.
4. Establishment of procedures to monitor the CCP's.
5. Establishment of the Corrective Action to be taken when monitoring shows that a CL has been exceeded.
6. Establishment of effective record keeping systems that document the HACCP system.
7. Establishment of procedures to verify that the HACCP system is working.

HACCP should be implemented NOW in every food plant by developing a plan that emulates your process with clearly marked critical control points followed by computerizing your data to monitor and provide corrective action instructions when and where needed.

HACCP is an effective system to focus on potential hazards during the process to assure the safety of the food. HACCP is cost effective and it reduces the need for testing of finished products to assure that the products being manufactured are safe.

HACCP can be most helpful and beneficial in today's marketing plan for any firm.

CHAPTER 23

MAINTENANCE, REPAIRS, AND PEOPLE

There is no question about the continual need for maintenance, and maintenance personnel in every food firm. From a contamination and food safety standpoint, the following suggestions may help prevent problems during repair, maintenance, and/or during construction.

1. All drill shavings and metal filings must be removed before startup.

2. All weld slag's welding rods, and spent rods must be removed before start-up.

3. All nuts, bolts, screws, gaskets, and repair plates must be picked up and returned to the storeroom.

4. Electrical components, wire, wirenuts, etc., must be picked up and removed from the area before start-up.

5. Excessive lubricants must be wiped off and all grease rags and towels must be picked up and removed before start-up.

6. All tools and instruments used in repairs must be returned to their appropriate areas before start-up.

7. All areas or equipment in repair must be cleaned before start-up.

8. Maintenance personnel must never allow their tools or equipment to come in contact with food belts, conveyors, tables or food contact services.

9. Maintenance personnel must be clean and wear appropriate head and hair restraints and, if needed, snoods while working in food areas.

10. Above all else, all precautions must be taken to assure the safety of the food during any breakdown, temporary repair, or maintenance work. Baffles, shields, diverters should be used as a precautionary measure.

11. Caulking materials, string, cardboard or wood, and bailing wire should never be used in food contact areas.

APPENDIX

The following industrial flow chart from Heat and Control® illustrates how the many unit operations go together to make a process for Corn chips.

The unit operations involved are shown by numbers on the opposite flow chart:

1. Corn And Water Mixing & Cooking
2. Soaking And Steeping to Loosen Pericarp
3. Washing to Remove Loosened Pericarp
4. Grinding, Extruding and Sheeting
5. Baking
6. Cooling and Equilibrating
7. Extruder For Chips
8. Fryer
9. Oil Filter
*. Oil Filter - Clean Sweep (not shown)
*. Alternate Coil Heat Exchanger (not shown)
*. Alternate Clean Heat Exchanger (not shown)
*. Alternate Control System (not shown)
*. Direct Heat Fryer (not shown)
*. Oil Tanks & Pumps (not Shown)
10. Oil Mist Eliminator
11. Salting
12. Cooling
13. Flavoring
14. Inspection
15. Incline Conveyor
16. Vibratory Distribution Feeder
17. Ishida Computer Combination Weigher
18. Packaging System Platform

---

* Not shown in the drawing on the next page

APPENDIX 167

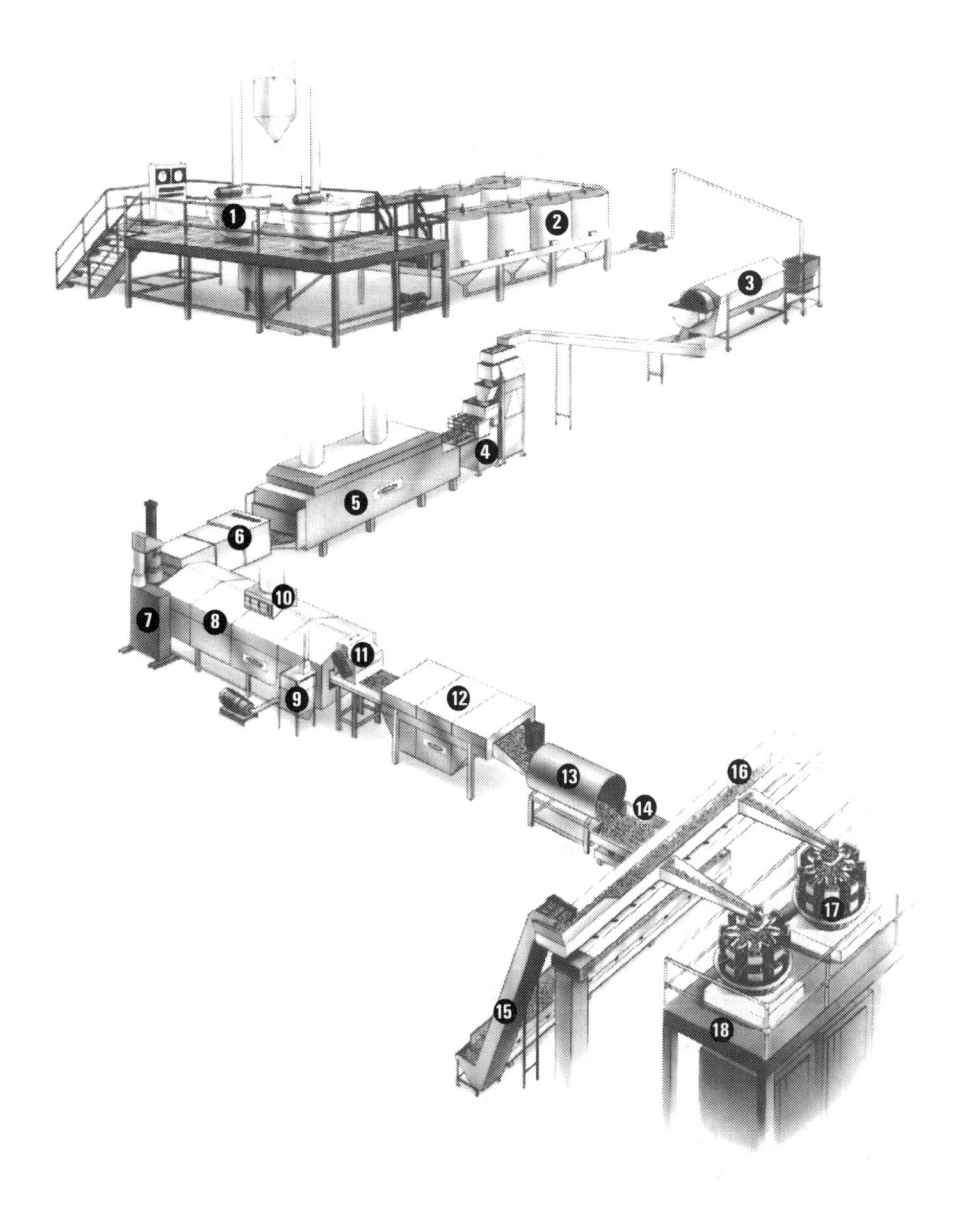

APPENDIX FIGURE 1 — Unit Operations In The Manufacture of Corn Chips (Courtesy Heat and Control)

REFERENCES AND FURTHER READINGS

Anon. 1950. "Steam Flow Closure." Research Bulletin. American Can Co., Inc. Maywood, IL.

Anon. 1982. *Thermal Processes for Low-acid Foods in Metal Containers.* Bulletin 26-L, 12th Ed. National Food Processors Association, Washington, DC.

Anon. 1988. *Canned Foods – Principles of Thermal Process Control, Acidification and Container Closure Evaluation.* Food Processors Institute, Washington, DC.

Anon. 1984. "How To Cut Food Products." Urschel Laboratories, Inc., Valparaiso, IL.

Anon. 1994. *Code of Federal Regulations, CFR 21 Parts 100-169.* Office of the Federal Register, National Archives and Records Administration, U.S. Government Printing Office, Washington, DC.

Anon. 1995. *The Almanac of the Canning, Freezing, Preserving Industries.* Edward E. Judge & Sons, Inc., Westminster, MD.

Boyd, J. M. and J. H. Bock. 1952. "Vacuum in Canned Foods – Its Significance and Its Measurement." Bulletin No. 31. Continental Can Co., Chicago, IL.

Boyd, J. M., J. M. Heinen, Jr. and F. W. Parrin. 1951. "Engineering the Air Removal from Canned Foods." Bulletin No. 29. Continental Can Co., Chicago, IL.

Brady, Robert C. 1952. "Material Handling. A Full New Science of Industrial Operation." Booklet No. 1, The Material Handling Institute, Pittsburgh, PA.

Buffington, M. A. 1969. "Mechanical Conveyors and Elevators." *Chemical Engineering.* Vol. 76 (22): 75-87. McGraw Hill Publishing Co., New York, NY.

Charm, S. E. 1978. *The Fundamentals of Food Engineering*, AVI Publishing Co., Inc., Westport, CT.

Cruess, W. V. 1958. *Commercial Fruit and Vegetable Products.* McGraw Hill Book Co., Inc., New York, NY.

Gould, Wilbur A., J. R. Geisman and J. P. Sleesman. 1959. "Washing Tomatoes." Research Bulletin 825. Ohio Agricultural Experiment Station, Wooster, OH.

Gould, Wilbur A. 1992. *Tomato Production, Processing & Technology,* CTI Publications, Baltimore, MD.

REFERENCES AND FURTHER READING

Gould, Wilbur A. 1994. *CGMP's/Food Plant Sanitation*. CTI Publications, Baltimore, MD.

Gould, Wilbur A. and Ronald W. Gould. 1993. *Total Quality Assurance for the Food Industries*. CTI Publications, Baltimore, MD.

Hall, Carl W. and D. C. Davis. 1979. *Processing Equipment for Agricultural Products*. AVI Publishing Co., Inc., Westport, CT.

Hardin, B. 1995. "Fruit Sorters Need More Light." USDA Agricultural Research, Vol. 43 (6): 19. ARS, USDA, Washington, DC.

Heldman, D. R. and R. P. Singh. 1982. *Food Process Engineering*. AVI Publishing Co., Inc., Westport, CT.

Holmquist, J. W., L. E. Clifcorn, D. G. Heberlein, C. F. Schmidt and E. C. Ritchell. 1954. "Steam Blanching of Peas." Continental Can Co., Inc., Chicago, IL.

Johanson, J. R. 1969. *Feeding*. Chemical Engineering Deskbook Edition. Vol. 6 (22): 7-105. McGraw Hill Publishing Co., New York, NY.

Nelson, P. E., J. V. Chambers and J. H. Rodriquez. 1987. *Principles of Aseptic Processing and Packaging*. The Food Processors Institute, Washington, DC.

Parker, M. E., E. S. Harvey and E. S. Stateler. 1952. *Elements of Food Engineering*. Vol. 1, 2 and 3. Reinhold Publishing Corp., New York, NY.

Perry, John H. 1950. *Chemical Engineers Handbook*. McGraw Hill Book Co., Inc., New York, NY.

Peterson, G. T. 1949. "Methods of Producing Vacuum in Cans." Bulletin No. 18. Continental Can Co. Inc., Chicago, IL.

Ratcliffe, J. D. 1975. "A Practical Analysis of the In-Plant Peeling Losses on Potatoes." Flo-Mech Limited, Whittlesey, Peterborough, England.

Suderman, D. R. and F. E. Cunningham. 1983. *Batter and Breading Technology*. AVI Publishing Co., Inc., Westport, CT.

INDEX

Aseptic Fillers 114

Bar Coding 124
Batter and Breading 101
Blanching, defined 75
    Methods 76
Blending, definition 89

Canning, history 1,3
    Principles 125
    Methods 128
Centrifugal Pumps 79,81
Chopping 60
Cleaning definition 27
Code and Coding 122
Comminuted 60
Container Closure, definition 119
Continuous Cooker/Cooler 131,136
Coring 52
Crateless Retort 130
Crushed 60
Cutting 58

Deaeration 66
Destemming 53
Disintegration 49,57
Dried or dehydrated, history 1,3
    Definition 143
    Methods 145
Double Seam 120
Dry Cleaning 27

Electronic Metal Detectors 71
Emulsified 62

Enrobing 101
Entoleters 70
Enzymes 75
Equipment requirement 7
Evaporator 135
Exhausting, definition 107
Extraction 63
Extrusion, definition 159
    Methods 160

Feeders 15
Filling and Fillers, definition 111
    Nomograph 112
    Filling Methods 112
Filtration and Filtering 65
Flow chart, generic 4
    Batter and Breading 102
Food additives 97
Food plant location, factors 5
Food Processing Industry Organization 2,3,5
Food Safety 163
Freezing, Definition 137
    History 1
    Methods 139
Frying, Definition 151
    Methods 154

HACCP 163
Hermetical Seal 119
Homogenized 61
Hot water blanchers and blanching 76
Husking 50

Industry magnitude 9

Juicing 63

Latent Heat of Foods 138
Liquid Material Handling 24

Magnetic Protectors 71

INDEX

Maintenance 165
Material Handling 11
Metal Detectors 71
Mixing, defined 89

Peeling, defined 41
Pellets 156
pH, definition 125,128
    Food Products 126,27
Pitting 52
Plate Heat Exchanger 133
Positive Displacement Pumps, ... 79,80
Protective equipment, in-line 69
Pumps and pumping, definition 79

Quality Separation, defined 33

Reasons for food processing 5
Receiving 12
Retort 129
Rotary Coils 134

Salt and Salting 97
Scraped Surface Heat Exchanger 134
Screens and Sieves 69
Seasoning 103
Separation, definition 65
Sheeting 64
Shelling 51
Shredding 60
Sifters 69
Size Separation 34
Snipping 53
Solid Material handling 19
Sorting 37
Specific Gravity Separation 37
Specific Heat of Foods 138
Steam blanchers and blanching 76
Steam flow closure 109
Stem-out 54
Storage bins 14

Sugar and syrups 99

Tubular Heat Exchanger 133

Unit Operations, defined 2

Vacuum 109
Vacuumizing, definition 107
Vertical Continuous Cooker 132
Vision Separation 36
Vitamins and other nutrients 105

Weight Separation 37
Wet Cleaning 30

INDEX

FIGURES

FIGURE 1.1	Generic Fruit and Vegetable Canning Flow Chart	4
FIGURE 2.1	Truck Lift for Unloading	13
FIGURE 2.2	Storage Bins	14
FIGURE 2.3	Funnel–Flow Bin Design is most common.	15
FIGURE 2.4	Mass–Flow Bins do not channel when material discharges.	16
FIGURE 2.5	Constant Pitch Screw Feeder gives this flow pattern.	16
FIGURE 2.6	Belt or Apron Feeder may produce dead spots in hopper.	16
FIGURE 2.7	Variable Pitch Screw gives uniform draw.	16
FIGURE 2.8	Tapered opening helps flow from belt or apron feeders.	17
FIGURE 2.9	Table Feeder has raised skirt.	17
FIGURE 2.10	Vibratory Feeders work well with slots.	18
FIGURE 2.11	Star Feeder is another device for uniform withdrawal from long slots.	18
FIGURE 2.12	Basic Conveyor, Flight and Pitch Types	20
FIGURE 2.13	Belt, Chain, Roller Conveyors	21
FIGURE 2.14	Drag Conveyor	21
FIGURE 2.15	Pivoted Bucket Conveyor	21
FIGURE 2.16	Pan Conveyor	22
FIGURE 2.17	Vibra-Free Distribution Conveyor	22
FIGURE 2.18	Flight Conveyor	23
FIGURE 2.19	Water or Flumes	23
FIGURE 2.20	Hydro Transfer System Series 7000	24
FIGURE 2.21	Even-Flow Storage and Surge Tank	25
FIGURE 3.1	Destoner	28
FIGURE 3.2	Air Blast Cleaner	29
FIGURE 3.3	Model DW-150 with Stone Crib	30
FIGURE 4.1	Rotary Size Grader	34
FIGURE 4.2	Vibra-Free Two-Deck Grader	35
FIGURE 4.3	Sortex 6000 Electronic Vision Sorter	36
FIGURE 4.4	USDA Dumping and Continuous Grading Belt	38
FIGURE 5.1	Batch Peeler	42
FIGURE 5.2	Continuous Brush and/or Abrasion Peeler	43
FIGURE 5.3	Lye Peeler/Scrubber/Washer for peeling tomatoes, other fruits and root crops.	44
FIGURE 5.4	'Saturno' Thermal Peeler	45
FIGURE 5.5	Vacuum Peeling Line	46
FIGURE 6.1	Pea Harvester (Courtesy Klockner Hamachek)	51
FIGURE 6.2	Snipper and Unsnipped Bean Remover (UBR) for Snap Beans	53
FIGURE 6.3	Stemout Machine Being Used On Tomatoes	54

FIGURE 7.1	8-Lane Sizer/Halver	58
FIGURE 7.2	Model CC Slicer	59
FIGURE 7.3	Tomato Chopper	60
FIGURE 7.4	Juice and Pulp Extractor	62
FIGURE 7.5	Belt Type Continuous Juice Extractor	63
FIGURE 9.1	Cleanline Metal Tracker System	71
FIGURE 9.2	Metal Tracker Pass-Through System	72
FIGURE 9.3	Metal Tracker Incline System	73
FIGURE 9.4	Metalchek 20	73
FIGURE 10.1	Rotary Hot Water Blancher/Cooker	77
FIGURE 11.1	Types of Pumps	80
FIGURE 11.2	Air-powered, Double-diaphram pump	85
FIGURE 11.3	Static Suction Head Levels for Three Products	86
FIGURE 11.4	Hydro Transport Food Pump	87
FIGURE 12.1	Jacketed Mixing Kettle	90
FIGURE 12.2	Jacketed Mixing Kettle Showing Agitators	91
FIGURE 12.3	Process Tank	91
FIGURE 12.4	Process Tank with Heat Exchange Surface	92
FIGURE 12.5	Process Tank with Agitation	92
FIGURE 12.6	Typical Batch and Continuous Processing Systems	93
FIGURE 12.7	Dry Mixing Installation and Storage Tank	94
FIGURE 13.1	Granular Salt Being Dispensed into Cans	98
FIGURE 13.2	Salt Tablet Dispenser	99
FIGURE 13.3	Frozen Block Processing System for Poultry or Fish	102
FIGURE 13.4	Breaded Products Fryer	103
FIGURE 13.5	Roll Salter	104
FIGURE 13.6	Tumble Drum Seasoner	105
FIGURE 15.1	With this nomograph and straight edge, you can compute dollar loss per hour for any situation.	112
FIGURE 15.2	Hand Pack Filler for Tomatoes	113
FIGURE 15.3	Rotary Piston Filler	113
FIGURE 15.4	Vacuum Filler	114
FIGURE 15.5	Belt Drive FFS Machine	115
FIGURE 15.6	Square Bottom Bag FFS Machine	116
FIGURE 15.7	Reciprocating Jaw Drive FFS Machine	116
FIGURE 15.8	Flo-Control Overflow Briner	117
FIGURE 16.1	Cross-Section of a Double Seam	120
FIGURE 16.2	Ink Jet Printer	123
FIGURE 16.3	Bar Code	124
FIGURE 17.1	Top: Vertical Retort, Bottom: Horizontal Retort	129
FIGURE 17.2	Crateless Continuous Retort System	130
FIGURE 17.3	Continuous Cooker/Cooler	131
FIGURE 17.4	Vertical Continuous Sterilizer	132
FIGURE 17.5	Plate Heat Exchanger	133
FIGURE 17.6	Tomato Juice Pasteurizer	133
FIGURE 17.7	Cut-away of a Scraped Surface Heat Exchanger	134
FIGURE 17.8	Rotary Coils for Concentration of Tomato Pulp	134

INDEX 177

FIGURE 17.9	Four-Stage Tomato Juice Evaporators	135
FIGURE 17.10	Pasteurizer Cooler	136
FIGURE 18.1	Batch Freezer	140
FIGURE 18.2	Double Contact Plate Freezer	141
FIGURE 18.3	Nitrogen Tunnel Freezer	142
FIGURE 19.1	Ideal Conditions within a Dehydrator under a Two-Stage Flow Dryer	145
FIGURE 19.2	Simple Concurrent Tunnel (Elevation)	146
FIGURE 19.3	Simple Counterflow Tunnel (Elevation)	146
FIGURE 19.4	Types of Drum Driers	147
FIGURE 19.5	Types of Spray Driers	149
FIGURE 20.1	Oil Treatment System	152
FIGURE 20.2	Pollution Control Heat Exchanger	153
FIGURE 20.3	Schematic of Pollution Control Exchanger	154
FIGURE 20.4	Batch Fryer	155
FIGURE 20.5	Continuous Formed Potato Products Fryer	155
FIGURE 20.6	Pellet Popper	156
FIGURE 20.7	Schematic of Heat Exchanger For Heating Wash Water or Specific Rooms	157
FIGURE 21.1	Cross-section of a Typical Cooking Food Extruder	160

TABLES

TABLE 1.1	Organization Chart For A Food Plant	6
TABLE 1.2	1992 Census Data For Specific SIC Numbers	8
TABLE 5.1	Relationship of Tuber Size to Various Physical Constants for Tubers of Different Diameters	47
TABLE 11.1	How Sanitary Rotary and Centrifugal Pumps Compare	84
TABLE 13.1	Relationship of Put-In (P-I) Syrup vs. Cut-Out (C-O) Syrup by Label Requirements for Standardized Fruit Products	100
TABLE 14.1	Comparison of Methods for Producing Vacuum in Canned Foods	109
TABLE 17.1	pH Values of Some Commercially Canned Foods	126
TABLE 18.1	Specific and Latent Heats of Foods	138

NOTES

NOTES

NOTES

NOTES

NOTES